Modern Perspectives in Otology

Advances in
Oto-Rhino-Laryngology

Vol. 31

Series Editor
C. R. Pfaltz, Basel

S. Karger · Basel · München · Paris · London · New York · Tokyo · Sydney

Symposium on Recent Advances in Otological Research,
in honour of H. F. Schuknecht, Boston, Mass., June 21 and 22, 1982

Modern Perspectives in Otology

Volume Editors
B. H. Colman, Oxford
C. R. Pfaltz, Basel

148 figures and 18 tables, 1983

S. Karger · Basel · München · Paris · London · New York · Tokyo · Sydney

Advances in Oto-Rhino-Laryngology

National Library of Medicine, Cataloging in Publication

Symposium on Recent Advances in Otological Research (1982: Boston, Mass.)
Modern perspectives in otology/
Symposium on Recent Advances in Otological Research, in honour of H. F. Schuknecht, Boston, Mass.,
June 21 and 22, 1982;
volume editors, B. H. Colman, C. R. Pfaltz – Basel; New York: Karger, 1983.
(Advances in oto-rhino-laryngology; v. 31)
1. Ear Diseases – congresses 2. Otolaryngology – congresses I. Colman, Bernard H. II. Pfaltz, C. R.
(Carl Rudolf) III. Schuknecht, Harold Frederick, 1917 – IV. Title V. Series
W1 AD701 v. 31 [WV 200 S989m 1982]
ISBN 3-8055-3641-0

Contents

Contents

Harold F. Schuknecht

Foreword

'It is my firm conviction that a knowledge of pathophysiology forms the basis for intelligent and successive prevention and management of disease. The source of such knowledge derives principally from laboratory research performed by individuals who are trained in scientific methodology.'

This statement can be found in the preface of *Harold F. Schuknecht's* monograph on 'pathology of the ear', and in a way it may be regarded as the 'credo' of a man who has dedicated his life both to basic science and to clinical work. *A. Flexner*, in his fundamental essay on medical education, has emphasized that neither in the nature of the thinking process nor in the scientific technique can a sharp line be drawn between practice and research in medicine. *Harold Schuknecht*, as a teacher, has always been conscious of this scientific approach to clinical training and this volume of our series shows very clearly the results of this educational principle: The selection of papers presented at the meeting of the HFS Society deal with animal studies, temporal bone pathology, ultrastructural and biochemical studies, clinical investigations and the results of applied clinical research. They reflect in the true sense of the word modern perspectives in otology on the one hand and on the other the scientific harvest of a great scientist, clinician and teacher.

Carl Rudolf Pfaltz

Preface

This volume contains a selection of papers presented at the meeting of the HFS Society in Boston on 21st and 22nd June, 1982. The membership of the HFS Society consists of those individuals who had the privilege of working as research fellows with Prof. *H. F. Schuknecht*, they share this common bond, and also the desire to know more about the ear in health and disease. The quality of the training and inspiration provided by Prof. *Schuknecht* is reflected in the fact that 11 of his former fellows are now full professors, and another 9 are department chairmen. The current membership is based on 77 former fellows from 25 different countries. The purpose of their meeting is to bring forward for discussion various research projects in which they have been involved in the preceding 3 years and the present volume is made up of a proportion of their papers presented at the 1982 meeting in Boston, together with a small number of guest papers.

The 1982 meeting was a special one in various ways. It coincided with the 200th Anniversary of the Harvard Medical School with which the Massachusetts Eye and Ear Infirmary is affiliated. It also coincided with the 21st Anniversary of the appointment of Dr. *Schuknecht* as LeComte Professor of Otology and Professor of Laryngology in the Harvard Medical School, and Chief of Otolaryngology at the Massachusetts Eye and Ear Infirmary. Although 1982 sees the retirement of him from the Chairmanship of the Department it will be good news to his friends and admirers throughout the world that he will be continuing working as professor.

This very short preface is not intended to be a eulogy to Prof. *Schuknecht*; suffice it to say that his scientific and clinical contributions in the field of otology are unlikely to be surpassed by anyone else in the 20th century and there will be few wishing to question this statement. This volume is dedicated to Dr. *Schuknecht* with affection and admiration by his present and former research fellows.

In the production of this volume I acknowledge with gratitude the continual helpfulness of the publishers, Messrs. S. Karger, and of Ms. *Denise Greder* in particular, as well as my colleagues *George Browning, Makoto Igarashi, Collin Karmody,* and *Alan Kerr* who assisted in the task of editing. I would like to thank Dr. *David Lim* for having negotiated with Messrs. Karger over the publication of these Proceedings. A particular debt of gratitude is due to Ms. *Carol Ota* who played such an important role in the organization of the meeting at every stage.

Bernard H. Colman

Adv. Oto-Rhino-Laryng., vol. 31, pp. 1–7 (Karger, Basel 1983)

Harvard Medical School and the Massachusetts Eye and Ear Infirmary

Harold F. Schuknecht

Department of Otology and Laryngology, Harvard Medical School, and
Department of Otolaryngology, Massachusetts Eye and Ear Infirmary,
Boston, Mass., USA

I am pleased that so many of our former research fellows have found it possible to return to Boston on the occasion of the third triennial meeting of the HFS Society. There are now 77 members of the Society from 25 countries and 61 of you are here today. A common bond, the wish to know more about the ear in health and disease, has brought you together to spend the next 5 days exchanging knowledge with each other. It has been an exciting experience for me to see how you, the members of this society, have continued to be productive researchers and have grown in academic leadership in your institutions. In the total membership there are now 11 full professors and 9 department chairmen, and there will be more.

Some of you had fellowships at the Henry Ford Hospital but most of you had your appointments here at the Harvard Medical School and the Massachusetts Eye and Ear Infirmary. All of us, fellows and faculty alike, are grateful for the opportunities that these institutions have provided for research and clinical experience in otology.

This year we celebrate the 200th anniversary of Harvard Medical School. I would like to review for you a little of the history of Harvard Medical School, the Massachusetts Eye and Ear Infirmary and some of the people who made these institutions what they are.

Harvard Medical School was founded in the fall of 1782 when *John Warren* was chosen Professor of Anatomy and Surgery and *Benjamin Waterhouse* was chosen Professor of Theory and Practice of Physic. Following the appointment of *Aaron Dexter* as Professor of Chemistry and Materia Medica in the following year, the school was ready to begin its work.

Fig. 1. The administrative building seen in this modern photograph is one of the seven original buildings of the marble quadrangle.

Waterhouse should be credited with bringing small pox vaccination to America. He knew of *Edward Jenner's* work in Gloucestershire, England. In 1799 *Waterhouse* published an article in Boston's *Columbian Sentinel* entitled 'Something Curious in the Medical Line' in which he described *Jenner's* proof that those who contracted cowpox had a slight illness, eruptions on the hands, and subsequent protection from small pox. *Waterhouse* wrote to *Jenner*, who sent him some vaccine which he used to inoculate his own children and several patients. Once recovery was complete, the patients were given 'matter' from small pox patients and, aside from the site of the vaccination, there were no signs of illness. It is ironic that small pox has now been eliminated from the face of the earth and small pox vaccinations are no longer required for travel to any part of the world.

Harvard Medical School's first class graduated on July 16th, 1788 with degrees of bachelor of medicine. It was not until 1811 that graduates were given the degree of doctor of medicine. Classes were first held in Cambridge in the basement of Harvard Hall which was located in Harvard Yard. There were several more moves including a site adjacent to the Bulfinch Building at the Massachusetts General Hospital (MGH), as well as a site now occupied by the new section of the Boston Public Library, before the school finally settled at its current location on Longwood Avenue. The marble quadrangle was dedicated in 1906 (fig. 1).

The original plans called for brick exteriors, but the contractor, the Norcross Company, offered to substitute marble from their quarry in Dorset, Vermont, without change from the contract price. One account has it that the origin of the marble at the quadrangle was the overlying inferior marble which had to be quarried first in order to reach better quality material which lay further in and was destined for the New York Public Library building on 5th Avenue and 42nd Street. *Charles William Elliot*, then President of Harvard University, is said to have declared that the quality of the marble and the circumstances under which it became available were not beneath Harvard's dignity. Should you have the opportunity to examine it, you will note that the white marble on the quadrangle contrasts with the elegant and mellow appearance of the superior marble existing on the New York Library facade.

Many distinguished physicians and surgeons have served the Harvard Medical School through the years. Seven generations of *Warrens* taught or studied at Harvard Medical School. One of them, *Joseph Warren*, was a General in the Continental Army and died in the Battle of Bunker Hill on the well-known date of June 17, 1775.

Oliver Wendell Holmes was a Professor of Anatomy and Dean of the School, having graduated from Harvard College in 1829. While history credits the Hungarian physician *Ignaz Semmelweis* with the discovery of the method of contagion of puerperal fever in 1846, there is good documentation that *Oliver Wendell Holmes* first made the observation 3 years earlier in 1843.

During its first century, Harvard Medical School was a proprietary institution managed by its own faculty. Fees were collected from the students and the proceeds were divided in lieu of salaries at the end of each year. In 1782 when Harvard Medical School was founded, there were no general hospitals in Massachusetts and everyone received their care at home.

The Bulfinch building of the MGH opened in 1821. In those days the Charles River passed by the front door and many patients were brought to the hospital by boat. As time passed, other hospitals joined the Harvard group of teaching hospitals, including the Boston City Hospital, Peter Bent Brigham Hospital, Beth Israel Hospital, Children's Hospital, Robert Breck Brigham Hospital, and The Deaconess Hospital.

On October 1, 1824, *Edward Reynolds* and *John Jeffries* opened a free eye clinic on the second floor of a building in Scollay Square, an area now known as Government Center.

Edwin Reynolds had spent a year training at the London Eye Infirmary when he returned to Boston in 1818 to find his father blinded by cataracts in both eyes. *Reynolds* removed his father's cataracts and his sight was restored. Word spread quickly and soon *Reynolds* had numerous eye patients appealing for his help.

From the very beginning ear patients also sought help in this clinic. In 1827, 3 years after its inception, the Infirmary was incorporated under the name of the Massachusetts Charitable Eye and Ear Infirmary. The first person to be trained in eye, ear, nose and throat at the Infirmary was *John Homer Dix* who served as House Officer from 1837 to 1840.

In 1846, *William Morton* gave his first public demonstration of the use of ether at the MGH and, only 6 months later, *John Homer Dix* performed the first eye surgery at the Massachusetts Eye and Ear Infirmary under ether anesthesia. The Infirmary was relocated in temporary quarters several times, when finally in 1850 it acquired a home of its own on Charles Street circle.

By 1870 the population of the Commonwealth of Massachusetts had grown to 1.5 million and the number of patients seeking assistance at the Massachusetts Eye and Ear Infirmary had grown to 5,000 per year. The archives of the Infirmary show that about one fourth of the patients suffered from diseases of the ear.

In those early years, the art of otology was poorly developed. These early physicians had much more success in treating eye cases and were reticent to become involved with otologic problems. Nonetheless, the patients were there and had to be cared for.

The first medical publication emanating from the Infirmary was authored by *Edward Jones Davenport*. The article appeared in the *Boston Medical and Surgical Journal* in 1837 and was entitled 'Polypi in the Meatus Auditorium externis'. The first person in Boston to dedicate himself to the practice of otology was *Edward Hammond Clarke*. For many years he gave a series of lectures on otology which were open to all and attended by some Harvard medical students.

> *Clarke* had a special arrangement with the Infirmary which allowed him to see his patients there and to use them for clinical demonstrations. He had to see his patients on a bright sunny day. To make the most of sunlight he employed a mirror attached to a stand by a universal joint for reflecting sunlight from a window into the room and then by means of a lens of about 2 inches in focal length, he directed the light through a silver speculum into the ear canal.

In 1870, the position of Aural Surgeon was created at the Infirmary and the appointment was given to *Clarence John Blake*. By this time *Edward Clarke* had moved up to become President of the Board of Managers. *Clarence Blake*, from the day of his appointment until the day of his retirement, a period of 35 years, was known as 'Mr. Otology' at the Infirmary.

In 1888 when Harvard Medical School added a compulsory fourth year to its course of study, otology was officially added to the curriculum. The

Board of Managers of the Infirmary responded by converting a former waiting room to a lecture hall and opening the clinic for teaching of Harvard medical students. Additionally, the surgeons requested and received from the Board of Managers the sum of $100.00 for an ear reference library. *Clarence Blake* was appointed Professor of Otology and Department Chairman. At about the same time, *John Orne Greene* was named Professor of Clinical Otology and became the first surgeon in America to perform the Wilde incision and the Schwarze simple mastoid operation.

Blake was well qualified, having studied for 3 years under the great Viennese otologist *Adam Politzer.*

Both *Clarence Blake* and *Orne Green* were true disciples of *Adam Politzer* and paid homage to their master by using the classical Latin terminology for ear diseases as was the practice of the great Austrian ear specialist. The medical records of those years contain such terms as: eczema pustulosum acutum (Impetigo), otitis catarrhalis secernens mucosa acuta cum hyperplasia tonsillae pharyngeae (acute otitis media and tonsillitis), otitis media suppurativa chronica cum carie mastoideae et abscessu cervicis (chronic otitis media, mastoiditis, and cervical abscess), otitis media insidiosa (otosclerosis), and surditas senilis (presbycusis).

The need for otological services grew rapidly so that 3 years later, in 1891, the Board of Managers found it necessary to purchase two dwellings adjacent to the Infirmary and convert them into a pavilion with 30 beds.

Clarence Blake had the knack for surrounding himself with talented young physicians, one of whom was *Frederick L. Jack*, who is remembered for performing the first stapedectomies in America in 1892.

In April of that year *Clarence Blake* injured his right hand and had to delegate much of his operative work to his assistant surgeon *Frederick L. Jack*. This gave *Jack* the opportunity to prove or disprove his previous belief that removal of the stapes could improve hearing. He had been removing the tympanic membrane, malleus and incus in the treatment of chronic otorrhea, and in three of these cases he had found it necessary to also remove the stapes. Whether the removal of the stapes in these cases was accidental or purposeful is questionable. Nonetheless, one or more of these patients had improved hearing. *Jack* requested permission to do more stapedectomies and *Blake* agreed. In the following 6 weeks he did 16 cases and a number of them apparently experienced improved hearing. In 1892 he reported his findings to the American Otological Society and a year later presented an additional 32 cases. By 1894 when he made his last report he had a total of 60 cases, but by then his enthusiasm had been lost. 4 years later *Gorham D. Bacon*, a friend of *Blake* and *Jack*, wrote in his textbook: 'the result following extraction of the stapes has been decided unsatisfactory and the operation has been condemned'. It seemed that *Jack's* bright hopes did not materialize. He did demonstrate that it was possible in some cases to remove the stapes without damage to the labyrinth, but it would be many years before the full significance of his observations would be appreciated.

In 1893, friends of the Infirmary established the Aural Surgeons Fund which originally consisted of 21 shares of American Bell Telephone stock valued at $4,179.00. The fund was used for the purpose of providing instruments, appliances and books for the use of the aural department. The Aural Surgeons Fund still exists today and is used for the same purposes.

From 1827 to 1870 the staff of the Massachusetts Eye and Ear Infirmary trained 21 house officers in the combined specialty of ophthalmology and otolaryngology, and between 1870 and 1900 another 35 were trained in ophthalmology and 12 in otology.

In 1899 the Infirmary moved into a building on the present site at 243 Charles Street. Although the Boston newspapers proclaimed in headlines that the Infirmary was a model facility having electric lighting, telephones and modern plumbing, the services by today's standards were modest. While waiting to be examined, the patients sat on uncomfortable hard wooden benches arranged along the walls of darkened clinic rooms. When admitted, they all shared large open wards. No private patients were treated at the Infirmary until 1915.

As the Infirmary became known to the residents of New England, the demand for services increased rapidly. As a matter of necessity, the otology service was also caring for many patients with disorders of the nose, throat and larynx. Thus, in 1904 the Board of Managers officially created the Department of Otolaryngology. It was not until 1932, however, that otolaryngology won the full recognition of Harvard Medical School with the creation of a professorship of laryngology. The chairman at Harvard Medical School therefore held two professorships, one in otology and one in laryngology. *Harris P. Mosher* was the first person to hold the combined appointments. When the American Board of Otolaryngology was established in 1925, *Mosher* became its first president, a position he held for 22 years.

While much clinical research had been performed and papers had been written by the ear, nose and throat physicians since the inception of the infirmary, laboratory research first became a reality in the middle 1950s with the establishment of laboratories of auditory physiology, virology, and biochemistry.

In 1974 when the Infirmary celebrated its 150th anniversary, the Department of Otolaryngology had served as a teaching base for the Harvard Medical School students for 88 years. This year the Harvard Medical School celebrates its 200th anniversary. It is recognized as one of the leading medical schools in the world and the 1981 Gorman Report has again designated Harvard Medical School as the top school in the nation.

Fig. 2. The tall square building just to the left of the center (arrow) is the Massachusetts Eye and Ear Infirmary. Clustered behind it are buildings of the Massachusetts General Hospital and in front of it is Charles Street and Storrow Drive. In the background slightly right of center is Government Center and in the far distance beyond the shipping channel is Logan International Airport.

The Massachusetts Eye and Ear Infirmary is fortunate to have a close affiliation with Harvard Medical School (fig. 2). Our greatly expanded modern building will provide an even better opportunity to discharge our responsibility as a teaching arm of the Harvard Medical School.

H. F. Schuknecht, PhD, Department of Otolaryngology, Massachusetts Eye and Ear Infirmary, 243 Charles Street, Boston, MA 02114 (USA)

Adv. Oto-Rhino-Laryng., vol. 31, pp. 8–17 (Karger, Basel 1983)

The Focal Points of Origin of the Cellular Components of Effusions in Seromucinous Otitis Media

Collin S. Karmody

Tufts University School of Medicine and The New England Medical Center, Boston, Mass., USA

There are two elements in effusions of the middle ear – fluid and cells. Both elements obviously originate from either the blood stream or from the mucosa of the middle ear and it would be expected that the entire mucosa contributes uniformly. Observations on the temporal bones of the squirrel monkey and humans, however, identify a number of discrete points in well-defined geographic areas that appear to be the predominant sites of production of the cellular elements. Surprisingly, the anatomic distribution of these sites is identical in the animal and human material.

Several Investigators [1–5] have reported extensively on the histologic, biochemical and immunologic changes in different types of otitis media. None of these authors, however, commented on the overall pattern of cellular activity in the mucosa of the middle ear. The possibility of differential cellular activity in the mucosa of the tympanum in chronic serous otitis media was first suggested on examination of human temporal bones. A study of the patterns of selective focal activity was, therefore, undertaken by light microscopic examination of histologic sections of the temporal bones of 10 young adult squirrel monkeys and 10 humans, giving a total of 40 bones.

Materials and Methods

The animals were from a group in which chronic seromucinous otitis media was induced by a technique described previously [6]. Serous effusions were first established by the injection of silicone around the cartilaginous part

of the Eustachian tube. The serous effusions were then converted to seromucinous effusions by the intratympanic injection of 0.1 ml or a 10⁴/ml suspension of *Diplococcus pneumoniae*. Animals were sacrificed at 2-week intervals beginning at 2 weeks after innoculation. Temporal bones were fixed by
intravital perfusion with 10% formalin and processed by the standard celloidin technique. Sections were cut at 20 μm and every tenth section was stained
with hematoxylin and eosin and mounted for light microscopy. Human temporal bones were processed by the same technique. Attempts to stain for
lymphocytes by the technique recently described for use with paraffin sections
were unsuccessful with the celloidain sections.

Of the ten pairs of human bones, nine were from infants and young children; one was from a 69-year-old adult. The cause of death of the human cases
varied widely. 4 of the children died from cardiopulmonary collapse second-
a ry to congenital anomalies, the other 5 children succumbed to other problems. The adult female died from a cerebrovascular catastrophe.

Histology

The most interesting observation was the comparative inactivity of a
substantial part of the mucosa of the middle ear. Generally the mucosal surface of the tympanic membrane and the mastoid air cell system made little or
no contribution to the cellular element of the effusion. Alternatively the areas
of the production of cells were confined to focal points in the protympanum,
on the medial wall of the middle ear, and the aditus ad antrum (fig. 1). These
areas were easily identified by their triangular configuration with apices toward the lumen of the middle ear (fig. 2). In the base of the triangle, deep in
tunica propria, a small dilated blood vessel, probably an arteriole, could be
identified. This vessel was surrounded and sometimes obscured by a dense
collection of mononucleated cells that collected on the lumen side of the vessel
(fig. 3). There was little cellular activity on the periosteal side. The collection
of cells extended in a pyramidal configuration through the tunica propria to
the surface epithelium. At this point the surface epithelium lost its definition
and was densely infiltrated by migrating cells (immunocytes) (fig. 3). Clusters
of cells could be seen in the adjacent lumen of the middle ear. Immediately
beyond the confines of these triangular hillocks, the mucosa was compara-
tively inactive. The tunica propria was slightly thickened but demonstrated
no evidence of cellular migration and the overlying epithelium was flat and
well ordered (fig. 3).

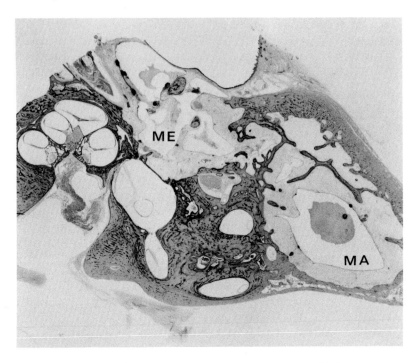

Fig.1. Horizontal section of the temporal bone of a 3-month-old infant. There are focal subepithelial collections of cells in the protympanum, promontory, and mastoid antrum. ME = Middle ear; MA = mastoid antrum. × 3.

Occasionally polypoid masses extended into the lumen of the middle ear, but none were seen in the mastoid cells. Polyps were defined as masses of tissue consisting of an extensive network of small blood vessels whose surface was covered by a single layer of flat cells. Polyps were a surprisingly active source of cells and were all surrounded by a dense zone of mostly mono-nucleated cells (fig. 4).

There were segments of the mucosa of the middle ear that seem to be un-involved or minimally involved in the inflammatory process. One such area was consistently seen on the surface of the horizontal canal and a second area on the anterosuperior segment of the tympanic membrane. In these areas the mucous membrane remained thin and flat. The subepithelial space was of normal thickness and the surface epithelium was normal (fig. 5). There was abrupt demarcation between these inactive areas and the rest of the mucosa of the middle ear.

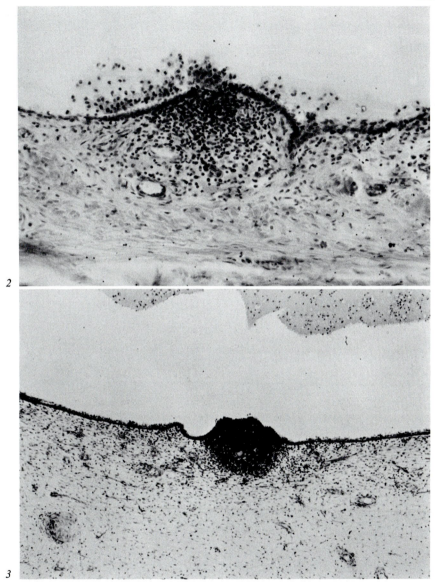

Fig.2. Focal collections of mononucleated cells migrating to a single point on the epithelium of the sinus tympani with extension through the epithelium. × 90.

Fig.3. Medium power microscopy of mucosa of the protympanum. A large number of immunocytes have collected on the lumen side of a small blood vessel. There is loss of definition of the surface epithelium while the surrounding epithelium and subepithelium tissues are relatively inactive. × 45.

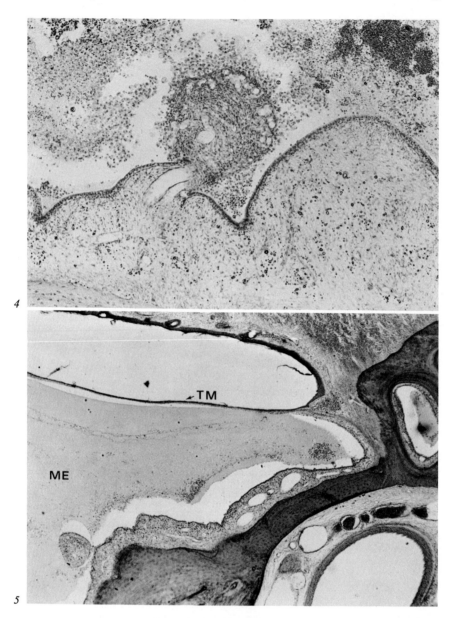

Fig. 4. Polypoid excrescence from the medial wall of the aditus ad antrum (animal). A dilated arteriole enters the base of the mass which consists of many dilated capillaries. Many mononucleated cells are migrating from the surface of the polyp. Note that the epithelium of the surrounding mucosa is intact. × 90.

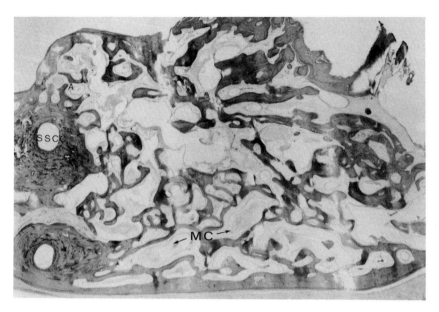

Fig.6. Mastoid process of a 69-year-old female. The mucosa of the cells is thickened but shows little cellular inflammatory reaction. SSCC = Superior semicircular canal; MC= mastoid cells; A = antrum. ×2,8.

There were active points on the medial wall of the mastoid antrum but much fewer foci on the lateral wall. The mucosa of the peripheral cells generally shows diminishing activity the further away it was from the mastoid antrum (fig. 6).

The lining of the protympanum was grossly thickened and thrown into folds – the surface epithelium was a high ciliated columnar type and there was increase in the population of secreting goblet cells. A careful search, however, identified no gland formation in the protympanum and the hypotympanum. This is contrary to the reports of *Arnold* [7], *Friedman* [8], and others. In 1 animal, on the anteroinferior promontory, there were intraepithelial microcysts

Fig.5. Section through the anterior middle ear of an experimental animal. The mucosal layer of the tympanic membrane (TM) is flat and shows no inflammatory reaction. In contrast the mucosa on the medial wall of the middle ear (ME) is thickened. In the tympanum is densely pink staining effusion with clumps of cells. ×7.

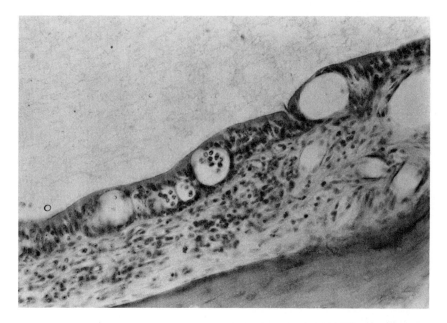

Fig. 7. Intraepithelial microcysts containing polymorphonuclear leukocytes. The surface of the epithelium is intact. × 179.

that contained a mixture of polymorphonuclear leukocytes, lymphocytes, and macrophages (fig. 7). The luminal wall of a few of the cysts were extremely thin suggesting imminent rupture, but the pattern indicated an unusually slow process of migration of polymorphonuclear leukocytes into the lumen of the middle ear. In one human specimen, one focus of perivascular cells consisted of plasma cells and eosinophils although the cellular component of this patient's middle ear effusion contained a few eosinophils.

Contrary to expectations, there were few foci of epithelial metaplasia. In the adult human specimen in two isolated areas of the mastoid antrum, the flat mucosa had changed to higher plumper goblet-like cells. There was, therefore, no evidence that, after 3 months, the mucosa of the middle ears of the squirrel monkey and man will change even focally to a squamoid keratin-producing type of epithelium which is contrary to the reports and concepts of other authors. The lack of metaplasia should be considered in the context of the composition of the epithelium of the cartilagonous part of the Eustachian tube as *Lim* et al. [1] reported that, for the most part, the floor of the cartilaginous Eustachian tube is lined by a simple squamous-like epithelium.

Discussion

It must be emphasized that all the material reported here were from ears with chronic seromucinous otitis media. The pattern in acute otitis media is different with a more uniform involvement of the mucosa of the middle ear.

Many investigators have studied the histopathology of all forms of otitis media, usually in biopsy material from patients, obtained during surgical maneuvers for persistent middle ear effusions. Most of the human biopsy material is taken from the promontory, from the region of the orifice of the Eustachian tube or from the hypotympanum. These are areas of intense activity and profuse production of cells that yielded much important information. Isolated minute biopsies, however, failed to identify the highly selective geographic distribution of the cell-producing areas. It is necessary, therefore, to obtain a panoramic view of the middle ear in total section for identification of the source of the cytologic element of otitis media.

Goycoolea et al. [9] studied experimentally induced acute and subacute otitis media in cats and found no lymphocytic cuffing of small vessels one month after obstruction of the Eustachian tube. *Ishii* et al. [10] detailed their observations in 13 temporal bones from 10 patients with otitis media with effusions. Eight ears had serous effusions and five had mucoid effusions. In three ears with mucoid effusions the mucosa over the promontory was covered with squamous or cuboidal epithelium. In five ears with serous effusions the submucosa was infiltrated with lymphocytes or neutrophils. In the ears with mucoid effusions, however, infiltration was rarely seen in the submucosa which is contrary to the findings in the specimens studied in this laboratory. These authors did find collections of neutrophils in the effusions close to subepithelial collections of neutrophils. In 1 74-year-old patient, supposed but not proven to have an immunologic disorder, the mucoid effusion contained lymphocytes and exfoliated epithelial cells.

Zechner [11] studied mucosal biopsies from patients and histologic sections of the temporal bones of humans with nonpurulent effusions in the middle ears. There were inflammatory changes in the mucosa and cellular infiltration limited to the proximity of small mucosal blood vessels.

Tos [12] and others have reported on the formation of neoglandular structures in the mucosa. Their material was primarily intraoperative biopsies. Although much discussion centered on the production of the fluid element of a middle ear effusion, there was little comment on the cell component. In the material being reported in this study, gland formation was sparse and was confined to the protympanum.

Veltri and Sprinkle [13] suggested that otitis media with effusion is caused by injury to the mucosa of the middle ear by complement-mediated immune complexes. They theorized that bacterial antigens from an antigen-antibody complex activate complement which in turn injures the basement membrane of the epithelium and stimulates chemotaxis of polymorphonuclear leukocytes. There is some experimental support for this hypothesis. If it is correct, however, the geographic selectivity for leukocyte chemotaxis still remains a mystery. The observations reported here suggest vulnerability of the mucosa at distinct points or alternatively these might be areas of enhanced immunologic responsiveness. It is also possible that the source of cells are transient points that are continuously shifting within the confines of a geographic area.

The cell-producing zones are somewhat closely related to the tracts of ciliated epithelium of the mucosa of the middle ear which are, in turn, areas of mucous secretion [14]. The nonciliated areas are much less active. Secretory cells of normal epithelium contain two distinct types of granules, mucous or dark granules [1]. The cells may contain only mucous granules or only dark granules or may be mixed. The dark cells may be protein-secreting cells and the mucous cells secrete a glycoprotein. The dark cells are probably related to the immunologic system of the middle ear. One can conceptualize that after release of the granules the dark cells collapse, thereby loosening the intercellular junctional areas through which the active immunocytes migrate to get to the lumen of the middle ear. There is no evidence in the material studied of a reversed migration of cells from the lumen back into the tunica, although there is a strong possibility that this does occur. It would be interesting to determine whether the same focal pathways are used in the reverse process.

In summary, horizontal sections of the temporal bones of squirrel monkeys and humans with chronic 'serous' otitis media (mucopurulent otitis) were studied. A well-defined pattern for the production of the cellular component of middle ear effusions was identified. The epithelial sites from which cells of chronic effusions migrate into the lumen of the middle ear are primarily in the protympanum, the promontory, and the epitympanum. The mucosa of the tympanic membrane and the mastoid cell system contribute only minimally.

References

1   Lim et al.: Normal and pathological mucosa of the middle ear and Eustachian tube. Clin. Otolaryngol. *4:* 213–234 (1979).

2   Senturia, B.H.; Carr, C.D.; Ahlvin, R.C.: Middle ear effusions. Pathological changes
    of the mucoperiosteum in the experimental animal. Ann. Otol. Rhinol. Lar. *71:*
    632–647 (1962).

3   Friedman, I.: The pathology of secretory otitis media. Proc. R. Soc. Med. *56:* 695–699
    (1963).

4   Tos, M.: Pathogenesis and pathology of chronic secretory otitis media. Ann. Otol.
    Rhinol. Lar. *89:* suppl. 68, pp. 91–97 (1980).

5   Bernstein, J.M.; Szymanski, C.; Albini, B.; Sun, M.: Lymphocyte subpopulations in
    otitis media with effusion. Pediat. Res. *12:* 766–768 (1978).

6   Rankin, J.D.; Karmody, C.S.: Serous otitis media, an experimental model. Archs
    Otolar. *92:* 14–23 (1970).

7   Arnold, W.: Reaktionsformen der Mittelohrschleimhaut. Archs Otolar. *216:* 369–473
    (1977).

8   Friedman, I.: The pathology of acute and chronic infection of the middle ear cleft.
    Ann. Otol. Rhinol. Lar. *80:* 390–396 (1971).

9   Goycoolea, M.M.; Paparella, M.M.; Juhn, S.K.; Carpenter, A.: Cells involved in
    the middle ear defense system. Ann. Otol. Rhinol. Lar. *89:* suppl. 68, pp. 121–218
    (1980).

10 Ishii, T.; Toriyama, M.; Suzuki, J.: Histopathological study of otitis media with
    effusion. Ann. Otol. Rhinol. Lar. *89:* suppl. 68, pp. 83–86 (1980).

11 Zechner, G.: Auditory tube and middle ear mucosa in nonpurulent otitis media. Ann.
    Otol. Rhinol. Lar. *89:* suppl. 68, pp. 87–90 (1970).

12 Tos, M.: Production of mucus in the middle ear and Eustachian tube. Ann. Otol.
    Rhinol. Lar. *83:* suppl. 11, pp. 44–58 (1974).

13 Veltri, R.W.; Sprinkle, P.M.: Secretory otitis media: an immune complex disease.
    Ann. Otol. Rhinol. Lar. *85:* suppl. 25, p. 135 (1976).

14 Shimada, T.; Lim, D.J.: Distributions of ciliated cells in the human middle ear. Ann.
    Otol. Rhinol. Lar. *81:* 203 (1972).

C.S. Karmody, MD, Tufts University School of Medicine and
The New England Medical Center, Boston, MA 02155 (USA)

Adv. Oto-Rhino-Laryng., vol. 31, pp. 18–27 (Karger, Basel 1983)

Uvulonodular Lesion and Eye-Head Coordination in Squirrel Monkeys [1]

M. Igarashi, H. Isago, T. O-Uchi, T. Kubo

Department of Otorhinolaryngology and Communicative Sciences,
Baylor College of Medicine, Houston, Tex., USA

Introduction

Electrical stimulation of the uvula and nodulus could elicit eye nystagmus [13, 49], and ablation studies showed various kinds of oculomotor abnormalities and the postural and locomotor imbalance [2, 16, 18, 21, 22, 24–26, 52, 53]. Furthermore, electrophysiological studies indicated that the uvula and nodulus receive not only the vestibular input [19, 20, 36, 44–46, 51, 59], but the visual input [11, 38, 45, 50, 51] and the neck proprioceptive input [47]. The multisensory convergences and efferent projections to the vestibular nuclei [3, 17, 47] suggest that these structures are involved in eye-head coordination. The coupled head and eye nystagmus could be evoked in the squirrel monkey, both by optokinetic and vestibular stimulations [32–34, 42], and the large negative correlation coefficient between slow phase velocities indicated that the normal animal could maintain a good eye-head coordination. The same head nonrestraint paradigm was used in this study in conjunction with uvulonodulectomy.

Methods

3 young healthy squirrel monkeys *(Saimiri sciureus)* were used. They were prescreened based upon having a good head movement. A light-weight pedestal with a vertical metal shaft was implanted on the monkey's skull. The

[1] This study was supported by McFadden Trust Research Fund and NINCDS Grant NS-10940.

head rotation in the yaw plane was recorded through a potentiometer, and the horizontal eye movements were recorded by a standard electronystagmography. All recordings were started 15 min after intramuscular injection of amphetamine sulfate (0.5 mg/kg).

The animal was placed on the turntable at a coaxial center of the 65 cm diameter white optokinetic drum with 16 vertical black stripes (1.7 cm wide and equally separated). Eye velocity was calibrated using a 30°/s optokinetic stimulus (with head restraint) at the beginning and end of each trial. The vestibular stimulus was 60 or 100°/s velocity step acceleration in the dark, and the optokinetic stimulus was 60 or 100°/s constant speed drum rotation. Stimulus directions were randomly mixed in order and the data were pooled. Recording was conducted with an interval of more than 7 days.

Upon acquisition of sufficient preoperative data, uvulonodulectomy was performed under general anesthesia [25, 26]. Postoperative recordings were continued for 6 months. Nystagmus samples for analysis were selected based on: head and eye nystagmus were coupled, constant, and were not interrupted. At the end of the experiment, animals were perfused and the brains and temporal bones were processed for light-microscopic investigation.

Results

In this experiment, the uvulonodulectomy did not produce directional dominance both in vestibulo- and optokinetic-oculomotor responses. Therefore, data from bidirectional stimulations in all monkeys were pooled.

Representative head and eye movements in a squirrel monkey evoked by 100°/s acceleratory velocity step (V-100) is shown in figure 1. The head rotated to the fast phase direction of eye nystagmus first, then gradually returned to the center position. Postoperatively, the amplitude of head rotation significantly decreased in both stimulus speeds (p < 0.05), and did not return to the preoperative level (fig. 2).

Superimposing on the head rotations, head nystagmus occurred, but the decay was faster than that of eye nystagmus. A negative correlation existed between slow phase velocities of coupled head and eye nystagmus. Slow phase head velocity (SPHV) and slow phase gaze velocity (SPGV) showed the maximum value immediately after the stimulus onset; however, slow phase eye velocity (SPEV) did not.

As shown in figure 3, postoperative maximum SPHV showed a significant decrease (p < 0.05), and did not recover. The SPEV contrarily showed

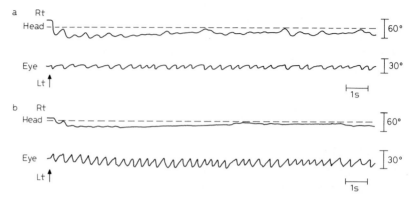

Fig. 1. Representative head and eye movements evoked by 100°/s acceleratory velocity step (arrows) before (a) and after (b) uvulonodulectomy. Horizontal dashed lines indicate the center position of head.

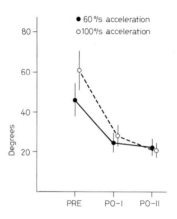

Fig. 2. Amplitude of head rotation (in degrees) provoked by vestibular stimulus. Vertical bars ± 2 standard error of means. PRE = preoperative stage; PO-I = early (0–10 weeks) postoperative stage; PO-II = late (10–24 weeks) postoperative stage (same for fig. 3, 4).

a very significant (approximately 100%) increase in the late postoperative stage. Those changes in SPHV and SPEV were similarly found by both stimulus speeds. Even though the sum of these did not show a significant change postoperatively, the characteristics of eye-head coordination obviously changed due to altered ratio between eye and head movements.

When the optokinetic stimulus was started, the head immediately rotated to the contralateral side of the stimulus direction and gradually returned to

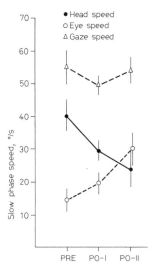

Fig. 3. Slow phase speeds (in °/s) of eye and head nystagmus evoked by 60°/s acceleratory velocity step, and the sum (gaze).

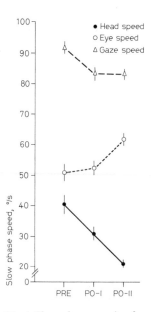

Fig.4. Slow phase speeds of eye and head nystagmus evoked by 100°/s optokinetic stimulus, and the gaze speed.

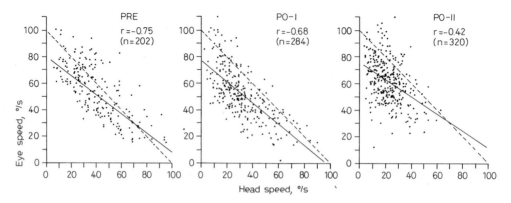

Fig.5. Three scattergraphs display coupled eye and head slow phase speeds evoked by 100°/s optokinetic stimulus. Note the postoperative reduction of caputomotor response and negative correlation coefficient (r).

the center position. Postoperatively, the amplitude of head rotation decreased significantly by both stimulus speeds, and failed to reattain the preoperative level.

Postoperative SPHV exhibited a very significant reduction ($p < 0.05$), whereas SPEV showed a significant increase only at the late postoperative stage (fig. 4) by both stimulus speeds ($p < 0.05$). SPGV showed a significant decline ($p < 0.05$), and never returned to the preoperative level.

Postoperatively, negative correlation coefficient between coupled head and eye speeds showed a significant reduction (more dominant in the late postoperative stage) in both stimulus speeds ($p < 0.05$) [12] (fig. 5). The light-microscopic evaluation of the neural tissues and temporal bones confirmed that the surgery did not involve any deep cerebellar structures, and middle and inner ear structures were all normal.

Discussion

Regarding the effect of uvulonodular lesions on vestibulo-oculomotor responses, *Dow* [16] reported no change in postrotatory nystagmus. *Westheimer and Blair* [57, 58], and *Blair and Gavin* [8] also showed no significant change in vestibulo-oculomotor reflex after complete or hemicerebellectomy. Contrarily, *Fernández and Fredrickson* [22] reported the hyperresponsiveness of vestibular nystagmus after uvulonodular operations. Also, *Grant* et al. [24] showed the postrotatory nystagmus asymmetry after unilateral lesion. All of these experiments were done under head restraint condition.

The results obtained in this study revealed increased vestibulo-oculo-motor response and decreased vestibulo-caputomotor response after the uvu-lonodulectomy. The former should be due to the reduction of cerebellar inhibitory control, because Purkinje cells in the vestibulo-cerebellum send inhibitory effects to the vestibular neurons which project to oculomotor neurons [4, 23, 28–30, 43], and the electrical stimulation of the nodulus produced a consistent inhibitory effect on vestibular nystagmus [22].

Neurophysiologically, it has been suggested that the vestibulo-cerebellum plays no role in the control of the vestibulospinal system [1, 27]. On the other hand, many lesion studies reported postablative changes [10, 16, 21, 22, 24, 25]. Also, there is an input from neck afferents to the cat flocculus [60, 61]. Moreover, the nodulus exerts a powerful inhibition on vestibular neurons transmitting vestibular and spinal input to the cerebellar nuclei and cortex [47]. These findings suggest that the vestibulo-cerebellum is linked to the vestibulo-spinal, vestibulo-cerebellar, and spino-cerebellar systems. The postablative decrease in vestibulo-caputomotor response (which failed to recover) in this study represents the effect from disruption of these neural connections, and verifies the importance of uvula and nodulus in this function.

The impairment of optokinetic nystagmus or pursuit eye movement after cerebellar lesions has been reported both in man and in animals under head-restraint conditions [9, 15, 37, 41, 54, 57, 58]. In squirrel monkeys, it was previously reported that the SPEV of optokinetic nystagmus reduced after uvulo-nodulectomy when the stimulus speed exceeded $90°/s$ [26]. Since the amplitude of eye saccades in head restraint condition and that of gaze in head unrestrained condition are identical [40], the SPGV reduction in this present study agrees with previous SPEV reduction under head restraint condition.

The stabilization of the eye during head rotation is from reflex action of some or all of the three feed-back circuits [6, 7], and the proper function of these systems will be indispensable for the exact eye-head coordination. Among the three oculomotor control systems, the cervico-oculomotor reflex has a relatively low gain [5, 39, 55, 56] and also it has a slow operation time [40]. Thus, this system plays only a limited role for the eye-head coordination. But a marked plasticity of this reflex loop gain after bilateral labyrinthectomy was reported. Thus, the postoperative changes in the vestibulo-caputomotor and optokinetic-caputomotor functions in the present study could secondarily affect this reflex system.

The optokinetic-oculomotor feedback loop has no functional role during saccadic gaze shift [40]; however, this system may play some role during the slow phase of nystagmus [32]. Thus, the postoperative increase of optokinetically evoked eye movement in this study could be one cause of the impaired eye-head coordination.

The vestibulo-oculomotor gain is close to unity and this reflex arc is not a feed-back system but a feed-forward system [14,48]. Also, the functional significance of vestibulo-oculomotor loop on eye-head coodination has been reported [14, 31, 34, 35]. Therefore, it is reasonable to consider that the post-lesion change of vestibular evoked eye movement in this present study is the main cause of the impaired eye-head coordination.

The multisensory convergence and connection to the vestibular nuclei indicate that the uvula and nodulus are involved in various oculomotor control loops and thereby related to eye-head coordination. The present experimental results in a primate model support this concept through the head non-restraint paradigm.

References

1   Akaike, T.; Fanardjian, V.V.; Ito, M.; Nakajima, H.: Cerebellar control of the vestibulospinal tract cells in rabbit. Exp. Brain Res. *18:* 446–463 (1973).
2   Allen, G.; Fernández, C.: Experimental observations in postural nystagmus. I. Extensive lesions in posterior vermis of the cerebellum. Acta oto-lar. *51:* 2–14 (1960).
3   Angaut, P.; Brodal, A.: The projection of the 'Vestibulocerebellum' onto the vestibular nuclei in the cat. Archs ital. Biol. *105:* 441–479 (1967).
4   Baker, R.; Precht, W.; Llinás, R.: Cerebellar modulatory action on the vestibulo-trochlear pathway in the cat. Exp. Brain Res. *15:* 364–385 (1972).
5   Barlow, D.E.; Freedman, W.: The cervico-ocular reflex in normal human adults. Soc. Neurosci. Abstr. *4:* 291 (1978).
6   Bizzi, E.: Motor coordination: central and peripheral control during eye-head movement; in Gazzaniga, Blakemore, Handbook of psychobiology, pp.427–437 (Academic Press, New York 1975).
7   Bizzi, E.; Kalil, R.E.; Morasso, P.; Tagliasco, V.: Central programming and peripheral feedback during eye-head coordination in monkeys. Biblthca ophthal., No.82, pp.220–232 (Karger, Basel 1972).
8   Blair, S.; Gavin, M.: Modification of the macaque's vestibulo-ocular reflex after ablation of the cerebellar vermis. Acta oto-lar. *88:* 235–243 (1979).
9   Burde, R.M.; Stroud, M.H.; Roper-Hall, G.; Wirth, F.P.; O'Leary, J.L.: Ocular motor dysfunction in total and hemicerebellectomized monkeys. Br. J. Ophthal. *59:* 560–565 (1975).
10 Carrea, R.M.E.; Mettler, F.A.: Physiologic consequences following extensive remo-

vals of the cerebellar cortex and deep cerebellar nuclei and effect of secondary cerebral ablations in the primate. J. comp. Neurol. *87:* 169–288 (1947).

11 Clarke, P.G.H.: The organization of visual processing in the pigeon cerebellum. J. Physiol., Lond. *243:* 267–285 (1974).

12 Cohen, J.; Cohen, P.: Applied multiple regression-correlation analysis for the behavioral sciences (Erlbaum Associates, Hillsdale 1975).

13 Cohen, B.; Goto, K.; Shanzer, S.; Weiss, A.H.: Eye movements induced by electric stimulation of the cerebellum in the alert cat. Expl. Neurol. *13:* 145–162 (1965).

14 Dichgans, J.; Bizzi, E.; Morasso, P.; Tagliasco, V.: Mechanisms underlying recovery of eye-head coordination following bilateral labyrinthectomy in monkeys. Exp. Brain Res. *18:* 548–562 (1973).

15 Dichgans, J.; Jung, R.: Oculomotor abnormalities due to cerebellar lesions; in Lennerstrand, Bach-y-Rita, Basic mechanisms of ocular motility and their clinical implications, pp.281–298 (Pergamon Press, Oxford 1975).

16 Dow, R.S.: Effect of lesions in the vestibular part of the cerebellum in primates. Archs Neurol. Psychiat., Lond. *40:* 500–520 (1938).

17 Dow, R.S.: Efferent connections of the flocculonodular lobe in *Macaca mulatta.* J. comp. Neurol. *68:* 297–305 (1938).

18 Dow, R.S.; Manni, E.: The relationship of the cerebellum to extraocular movements; in Bender, The oculomotor system, pp.280–292 (Harper & Row, New York 1964).

19 Ferin, M.; Grigorian, R.A.; Strata, P.: Purkinje cell activation by stimulation of the labyrinth. Pflügers Arch. ges. Physiol. *321:* 253–258 (1970).

20 Ferin, M.; Grigorian, R.A.; Strata, P.: Mossy and climbing fibre activation in the cat cerebellum by stimulation of the labyrinth. Exp. Brain Res. *12:* 1–17 (1971).

21 Fernández, C.: Interrelations between flocculonodular lobe and vestibular system; in Rasmussen, Windle, Neural mechanisms of the auditory and vestibular systems, pp.285–296 (Thomas, Springfield 1960).

22 Fernández, C.; Fredrickson, J.M.: Experimental cerebellar lesions and their effect on vestibular function. Acta oto-lar. *192:* suppl., pp.52–62 (1964).

23 Fukuda, J.; Highstein, S.M.; Ito, M.: Cerebellar inhibitory control of the vestibuloocular reflex investigated in rabbit. IIIrd nucleus. Exp. Brain Res. *14:* 511–526 (1972).

24 Grant, G.; Aschan, G.; Ekvall, L.: Nystagmus produced by localized cerebellar lesions. Acta oto-lar. *192:* suppl., pp.78–84 (1964).

25 Igarashi, M.; Miyata, H.; Alford, B.R.; Wright, W.K.: Experimental cerebellar uvulonodular lesions in the squirrel monkey. Adv. Oto-Rhino-Laryng., vol.19, pp.220–231 (Karger, Basel 1973).

26 Igarashi, M.; Miyata, H.; Kato, Y.; Wright, W.K.; Levy, J.K.: Optokinetic nystagmus after cerebellar uvulonodulectomy in squirrel monkeys. Acta oto-lar. *80:* 180–184 (1975).

27 Ito, M.: Neural design of the cerebellar motor control system. Brain Res. *40:* 81–84 (1972).

28 Ito, M.; Nishimaru, N.; Yamamoto, M.: The neural pathways mediating reflex contraction of extraocular muscles during semicircular canal stimulation in rabbits. Brain Res. *55:* 183–188 (1973).

29 Ito, M.; Nishimaru, N.; Yamamoto, M.: The neural pathways relaying reflex inhibition from semicircular canals to extraocular muscles of rabbits. Brain Res. *55:* 189–193 (1973).

30   Ito, M.; Nishimaru, N.; Yamamoto, M.: Specific neural connections for the cerebellar control of vestibulo-ocular reflexes. Brain Res. *60:* 238–243 (1973).

31   Kasai, T.; Zee, D.S.: Eye-head coordination in labyrinthine-defective human beings. Brain Res. *144:* 123–141 (1978).

32   Kubo, T.; Igarashi, M.; Jensen, D.W.; Homick, J.L.: Eye-head coordination during optokinetic stimulation in squirrel monkeys. Ann. Otol. Rhinol. Lar. *90:* 85–88 (1981).

33   Kubo, T.; Igarashi, M.; Jensen, D.W.; Wright, W.K.: Eye-head coordination and lateral canal block in squirrel monkeys. Ann. Otol. Rhinol. Lar. *90:* 154–157 (1981).

34   Kubo, T.; Igarashi, M.; Jensen, D.W.; Wright, W.K.: Head and eye movements following vestibular stimulus in squirrel monkeys. ORL *43:* 26–38 (1981).

35   Lanman, J.; Bizzi, E.; Allum, J.: The coordination of eye and head movement during smooth pursuit. Brain Res. *153:* 39–53 (1978).

36   Llinás, R.; Precht, W.; Clarke, M.: Cerebellar Purkinje cell responses to physiological stimulation of the vestibular system in the frog. Exp. Brain Res. *13:* 408–431 (1971).

37   Mackay, W.A.; Murphy, J.T.: Cerebellar modulation of reflex gain. Prog. Neurobiol. *13:* 361–417 (1979).

38   Maekawa, K.; Simpson, J.I.: Climbing fiber responses evoked in vestibulocerebellum of rabbit from visual system. J. Neurophysiol. *36:* 649–666 (1973).

39   Meiry, J.L.: Vestibular and proprioceptive stabilization of eye movements; in Bach-y-Rita, Collins, The control of eye movements, pp.483–496 (Academic Press, New York 1971).

40   Morasso, P.; Bizzi, E.; Dichgans, J.: Adjustment of saccade characteristics during head movements. Exp. Brain Res. *16:* 492–500 (1973).

41   Nemet, P.; Ron, S.: Cerebellar role in smooth pursuit movement. Documenta ophth. *43:* 101–107 (1977).

42   O-Uchi, T.; Igarashi, M.; Kubo, T.: Effect of frontal-eye-field lesions on eye-head coordination in squirrel monkeys. Ann. N.Y. Acad. Sci *374:* 656–673 (1981).

43   Precht, W.: Vestibular and cerebellar control of oculomotor functions. Biblthca ophthal., No.82, pp.71–88 (Karger, Basel 1972).

44   Precht, W.; Llinás, R.: Functional organization of the vestibular afferents to the cerebellar cortex of frog and cat. Exp. Brain Res. *9:* 30–52 (1969).

45   Precht, W.; Simpson, J.I.; Llinás, R.: Responses of Purkinje cells in rabbit nodulus and uvula to natural vestibular and visual stimuli. Pflügers Arch. *367:* 1–6 (1976).

46   Precht, W.; Volkind, R.; Blanks, R.H.I.: Functional organization of the vestibular input to the anterior and posterior cerebellar vermis of cat. Exp. Brain Res. *27:* 143–160 (1977).

47   Precht, W.; Volkind, R.; Maeda, M.; Giretti, M.L.: The effects of stimulating the cerebellar nodulus in the cat on the responses of vestibular neurons. Neuroscience *1:* 301–312 (1976).

48   Robinson, D.A.: Adaptive gain control of vestibuloocular reflex by the cerebellum. J. Neurophysiol. *39:* 954–969 (1976).

49   Ron, S.; Robinson, D.A.: Eye movements evoked by cerebellar stimulation in the alert monkey. J. Neurophysiol. *36:* 1004–1022 (1973).

50   Simpson, J.I.; Alley, K.E.: Visual climbing fiber input to rabbit vestibulo-cerebellum: a source of direction-specific information. Brain Res. *82:* 302–308 (1974).

51   Simpson, J.I.; Precht, W.; Llinás, R.: Sensory separation in climbing and mossy fiber inputs to cat vestibulocerebellum. Pflügers Arch. *351:* 183–193 (1974).

52 Spiegel, E.A.; Scala, N.P.: Vertical nystagmus following lesions of the cerebellar vermis. Archs Ophthal., N.Y. *26:* 661–669 (1941).
53 Spiegel, E.A.; Scala, N.P.: Positional nystagmus in cerebellar lesions. J. Neurophysiol. *5:* 247–260 (1942).
54 Takemori, S.; Cohen, B.: Loss of visual suppression of vestibular nystagmus after flocculus lesions. Brain Res. *72:* 213–224 (1974).
55 Takemori, S.; Suzuki, J.: Eye deviations from neck torsion in humans. Ann. Otol. Rhinol. Lar. *80:* 439–444 (1971).
56 Thoden, U.; Wirbitzky, J.: Influence of passive neck movements on eye position and brainstem neurons. Pflügers Arch. *362:* R37 (1976).
57 Westheimer, G.; Blair, S.M.: Oculomotor defects in cerebellectomized monkeys. Investve Ophth. *12:* 618–621 (1973).
58 Westheimer, G.; Blair, S.M.: Functional organization of primate oculomotor system revealed by cerebellectomy. Exp. Brain Res. *21:* 463–472 (1974).
59 Wilson, V.J.; Anderson, J.A.; Felix, D.: Unit and field potential activity evoked in the pigeon vestibulocerebellum by stimulation of individual semicircular canals. Exp. Brain Res. *19:* 142–157 (1974).
60 Wilson, V.J.; Maeda, M.; Franck, J.I.: Input from neck afferents to the cat flocculus. Brain Res. *89:* 133–138 (1975).
61 Wilson, V.J.; Maeda, M.; Franck, J.I.: Inhibitory interaction between labyrinthine, visual and neck inputs to the cat flocculus. Brain Res. *96:* 357–360 (1975).

M. Igarashi, MD, Department of Otorhinolaryngology and Communicative Sciences, Baylor College of Medicine, Houston, TX 77030 (USA)

Adv. Oto-Rhino-Laryng., vol. 31, pp. 28–38 (Karger, Basel 1983)

Morphological Study of the Middle and Inner Ear of the Bat: *Myotis myotis*

L. Cabezudo, F. Antoli-Candela, Jr., J. Slocker

Centro Especial Ramon y Cajal, Madrid, Spain

Introduction

The importance of the auditory system in the life of the bat has been recognized for almost two centuries. In 1794, *Spallanzani* [21] proved that blindness did not change the normal behavior of these mammals but that they were very much affected when their ears were plugged. Later experiments [4–6] have also shown that their cochleas are sensitive to ultrasonic frequencies. Much work has been done since to identify the morphological bases of echolocation [2,9,11,14,15,17]. The present paper offers further details on the structure of the bat's peripheral auditory system, placing special emphasis on those features possibly related to their ability to perceive high frequencies. The *Myotis myotis* was chosen as the experimental animal because it is a chiropter commonly found in Spain and its morphological characteristics were hithero practically unknown.

Material and Methods

Eight ears from four *Myotis myotis* have been studied. The temporal bones were intravitally perfused with Heidenhain-Susa, embedded in celloidin [20] and serially sectioned in the horizontal plane at 20 μm thickness. With the exception of two specimens in which all of the sections were used, every third section was stained in haematoxylin-eosin and examined under the light microscope.

The cochleae were reconstructed [7, 19] and the average length, 6.25 mm, was normalized to 100% (fig. 1a). Measurements of the cochlear width and height were carried out in the midmodiolar plane. The radial widths of the basilar membrane and spiral laminas (fig. 1b) were established as between 35 and 85% of the cochlear length. The areas of the round window membrane, oval window and spiral ligament were measured. The former was obtained with a polar planimeter by projecting the fastening points on the plane of the basilar membrane (fig. 1c) and the second by adding the partial areas in each serial section. The surface of the spiral ligament was calculated by the projection of its outline on a calibrated paper.

Results

The *Myotis myotis* middle ear consists of a simple air-filled cavity which communicates with the pharynx through the Eustachian tube. The medial part of the floor is membranous and cartilaginous, since it is not completely enclosed by bone. Pneumatization of the surrounding structures does not occur.

The tympanic cavity is transversed by three ossicles: the malleus, the incus and the stapes. The head of the *malleus* is joined to the incus and its long neck has two processes, lateral and anterior, whereas the manubrium is embedded in the tympanic membrane. The anterior process is very long and forms an osseous ankylosis with the tympanic bone (fig. 2a). The *incus* is composed of a small body and two other processes of differing size. A short pedicule interconnects the long process to the lenticular apophysis. The *stapes* has a head and two crura that are joined to the footplate by a cartilaginous tissue. The stapedial artery, a branch of the internal carotid artery, persists in the adult animal and passes through the intercrural arch on its way towards the facial nerve (fig. 2b).

In the middle ear there are two very well-developed muscles (fig. 2): the *tensor tympani* and the *stapedius*. The fibers of the former are arranged in tight bundles, with fat cells being occasionally seen among them. It is inserted in the malleus by a single tendon. The stapedius is located in the stapedial fossa adjacent to the third portion of the facial nerve. Its fibers originate in the walls of the fossa and extend on to a small bone or cartilage embedded within the muscle that is commonly known as os quartum or cartilage of Paaw [12]. From this point a tendon runs out to the stapes.

The ossicular chain also has two strong ligaments; the annular ligament

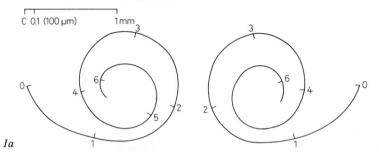

1a

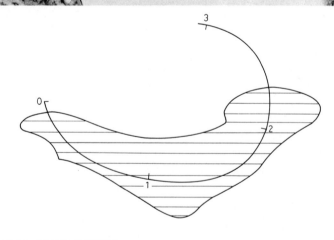

1b

1c

which firmly holds the footplate to the margins of the oval window and the posterior incudal ligament whose thickness reached 0.22 mm. Calcium deposits are often seen among the fibers at its insertion in the fossa incudis.

The cochlea of the *Myotis myotis* does not have any firm union to the neighboring sphenoid and occipital bones, but is separated from them by wide fissures of connective tissue crossed by large vessels (fig. 3a). The bony cochlea has an axial length of 2.20 ± 0.11 mm and is shaped into a spiral canal about 5.90–6.75 mm in length which winds 2¼ turns around the modiolus. The widths of the basal and middle turns are 2.70 ± 0.82 and 1.80 ± 0.73 mm, respectively. The walls are thin and compact with some areas of cancellous bone at the apex.

The modiolus has a hollow center to house the cochlear division of the eight nerve. It has an anterior direction and is practically parallel to the opposite side. The 3 turns therefore protrude into the cavity of the middle ear. Both spiral laminas are very well developed (fig. 1b). The primary shows a considerable thickening that stands out in the vestibular surface and the secondary is evident all along the cochlea. Both laminas are much more pronounced in the basal turn than in the apex, the primary being about double in size and the secondary is almost five times the size basally (table I, fig. 4).

The perilymphatic spaces are quite large, especially the scala tympani in the basal turn. At this level the perilymph communicates directly with the cerebrospinal fluid through the cochlear aqueduct (fig. 3b), which has a mean width of 159 ± 26 μm and a length of 794 ± 74 μm. The round window membrane traces the separation between the basal end of the scala tympani and the middle ear. It is located slightly anterior and inferiorly (fig. 3) and is not only attached to the margins of the round window but also reaches the base of the secondary spiral lamina and partially surrounds the modiolus. Its extension is 1.92 ± 0.23 mm². The ratio of the oval window to the round window membrane area is 1–14, since the surface of the former is 137 ± 0.11 μm².

The cochlear duct extends between both spiral laminas (fig. 1b). Its morphological characteristics are similar to those described in the rest of the mammals with the exception that it does not form a hook at the basal end. The

---

Fig. 1. a Graphic reconstruction of the cochlea. *b* Cochlear duct in the upper middle turn. The Böttcher cells can be seen at this level (▲). The arrows indicate the reference points taken for the measurements of the width of the basilar membrane (BM), pars pectinata (pp), primary spiral lamina (PSL) and secondary spiral lamina (SSL). The double arrows show the vestibular thickening. SL = Spiral ligament. *c* Area of the round window membrane and its relation to the reconstructed cochlear duct.

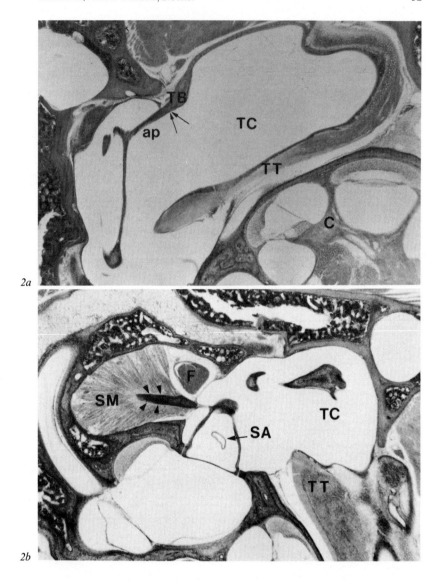

Fig. 2. Middle ear. TC = Tympanic cavity; TT = tensor tympani muscle. *a* The arrows point out the osseous ankylosis between the anterior process of the malleus (ap) and the tympanic bone (TB). C = Cochlea. *b* SM = Stapedius muscle and os quartum (▲) F = facial nerve; SA = stapedial artery.

Fig. 3. Cochlea. M = Modiolus; ST = scala tympani. The arrows show the round window membrane. *a* CT = Connective tissue. *b* CA = Cochlear aqueduct; PF = posterior fossa. Note the continuity of the malleus with the tympanic bone (↑↑).

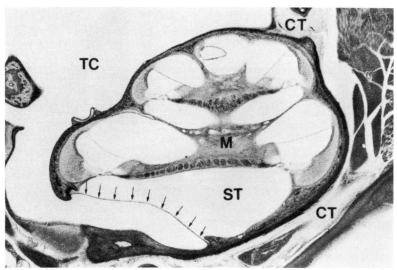

3a

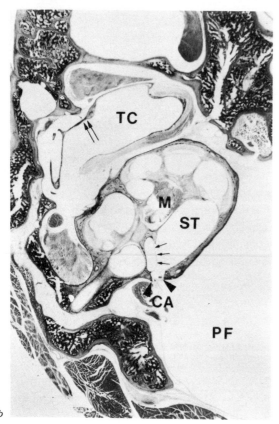

3b

Table I. Cochlear measurements (n = 8)

	Approximate location, mm							
	2		4		5		6	
	M	S	M	S	M	S	M	S
Width, μm								
Primary spiral lamina	406.25	22.79	285.00	11.64	288.12	18.69	185.00	21.17
Secondary spiral lamina	181.25	13.29	103.12	6.09	68.12	9.97	25.31	2.81
Basilar membrane	123.37	4.31	136.25	3.53	147.18	8.90	139.06	7.55
Pars pectinata	69.06	3.51	78.59	2.94	85.62	4.17	81.87	7.64
Pars tecta	54.31	1.13	57.66	0.82	61.56	6.68	57.19	0.12
Area, μm²								
Spiral ligament	614.00	104.00	308.00	47.40	285.00	46.90	124.00	19.80

M = Mean; S = standard deviation.

area of the spiral ligament reduces to less than one seventh its size from base to apex whereas the width of the basilar membrane increases (apart from a slight decrease which is found in the last part of the apical turn) (table I, fig. 4). The maximum width was 147.18 ± 8.90 μm and the minimum was 123.37 ± 4.31 μm, which represents an increase of only 1.20 times from 35 to 83.5% of the cochlear length. The pars tecta and the pars pectinata seem to contribute equally to this increment (table I).

The organ of Corti has the same structural elements as in other mammals. The distribution is also similar, with the exception of the fact that the Böttcher cells can be found up to the middle turn (fig. 1b) and that the outer hair cells are arranged in two rows at the apical end of the cochlea (fig. 5).

Discussion

Bats produce high frequency sounds in the form of short pulses which allow them not only to identify the existence of any surrounding object, but also its characteristics [5, 8, 13]. Several families of bats have been previously analyzed in order to establish a correlation between the structure of the auditory apparatus and these sounds [2, 9, 11, 14, 15, 17].

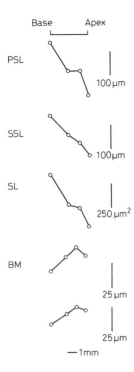

Fig.4. Average width of the primary spiral lamina (PSL), secondary spiral lamina (SSL) and basilar membrane (upper trace in BM) including the pars pectinata (lower trace in BM). Area of the spiral ligament (SL).

The present study shows that the middle ear of the *Myotis myotis* has a small ossicular mass firmly attached to the walls of the tympanic cavity by the continuity of the malleus with the tympanic bone and by the two strong ligaments, annular and posterior incudal ligament which form a structural design that would seem specifically aimed at the transmission of high frequencies.

The muscles of the middle ear appear to perform both protective and analytical functions. By contracting in the presence of loud sounds the transfer of vibratory energy to the cochlea is attenuated. *Henson* [11] has estimated that it reaches 30 dB for frequencies up to 50 kHz. This prevents overstimulation and may serve to keep the sound energy within the linear response range of the ear [10].

The isolation of the cochlea from the surrounding bones has an unknown meaning. One possible explanation might be that it is a protective mechanism

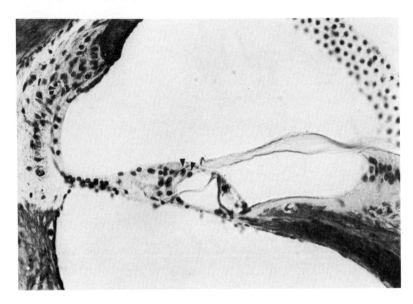

Fig. 5. Organ of Corti in the apex. At this level the outer hair cells are arranged in two rows (▲).

to lessen the osseous transmission of the vibration of the vocal cords during the emission of pulses. However, they should also be conducted from the cerebrospinal fluid to the perilymph through the wide cochlear aqueduct. The large extension of the round window membrane (1.92 ± 0.23 mm^2) is related to the great volume of both scalae in the basal turn.

Neither the histological cochlear study of the *Myotis myotis* nor those done in different species of bats [2, 11, 14, 15], have brought about any conclusive evidence of a specific structural pattern of the organ of Corti. Some morphological characteristics are found to be much more conspicuous than in other mammals, for example the great size of the spiral ligament, the existence of a long secondary spiral lamina that extends to the apex and the unusual thickness of the primary spiral lamina. Similar findings have been reported in aquatic animals with sonar skills [3, 16]. The peculiar development of these structures would necessarily increase the inflexibility of the vibratory system, since all of them take part in the anchorage of the basilar membrane and move when stimulated [1, 18].

The importance of the basilar membrane in the frequency selectivity of the cochlea has been widely recognized. Its width, thickness and other morphological characteristics largely determine the vibratory pattern along the cochlear duct. The basilar membrane of the *Myotis myotis* shows a slight increase (approximately 1.20-fold) between the basal and the upper middle turn. These measurements agree in general with those obtained in previous studies in other bats [2,9,11,14,15,17]. The coexistence of a narrow basilar membrane attached to the walls of well-developed spiral laminas has to offer great resistance to displacement. These structural characteristics must be related to its sensitivity to very high frequencies.

References

1 Bekesy, G. von: Experiments in hearing (McGraw-Hill, New York 1960).
2 Bruns, V.: Basilar membrane and its anchoring system in the cochlea of the greater horseshoe bat. Anat. Embryol. *161:* 29–50 (1980).
3 Fleischer, G.: On bony microstructures in the dolphin cochlea, related to hearing. N. Jb. Geol. Paleont. Abh. *151:* 166–191 (1976).
4 Galambos, R.: Cochlear potentials elicited from bats by supersonic sounds. J acoust. Soc. Am. *14:* 41–49 (1942).
5 Griffin, D.R.: Listening in the dark (Yale University Press, New Haven 1958).
6 Grinnell, A.D.: Neurophysiological correlates of echo-location in bats; PhD thesis and Technical Report No.30, Office of Naval Research, Cambridge (1962).
7 Guild, S.R.: A graphic reconstruction method for the study of the organ of Corti. Anat. Rec. *22:* 141–157 (1921).
8 Hartridge, H.: The avoidance of objects by bats in their flight. J. Physiol., Lond. *54:* 54–57 (1920).
9 Henson, M.M.: The basilar membrane of the bat, *Pteronotus p.parnellii.* Am. J. Anat. *153:* 143–158 (1978).
10 Henson, O.W.: The activity and function of the middle ear muscles in echolocating bats. J. Physiol., Lond. *180:* 871–887 (1965).
11 Henson, O.W.: The ear and audition; in Wimsatt, Biology of bats II, pp.181–262 Academic Press, New York 1970).
12 Klaauw, C.J. van der: Die Skelettstücken in der Sehne des *Musculus stapedius* und nahe dem Ursprung der Chorda tympani. Z. Anat. EntwGesch. *69:* 32–83 (1923).
13 Pierce, G.M.; Griffin, D.R.: An experimental determination of the supersonic notes emitted by bats. J. Mammal. *19:* 454–455 (1938).
14 Pye, A.: The structure of the cochlea in Chiroptera. I. Microchiroptera: Emballon-uroidea and Rhinolophoidea. J. Morph. *118:* 495–510 (1966).
15 Pye, A.: The structure of the cochlea in Chiroptera. II. The Megachiroptera and Ves-pertilionoidea of the Microchiroptera. J. Morph. *119:* 101–120 (1966).
16 Ramprashad, F.; Corey, S.; Ronald, K.: The anatomy of the seal's ear (*Pagophilus groenlandicus,* Erxleben, 1777); in Harrison, Functional anatomy of marine mammals 1, pp.264–305 (Academic Press, New York 1972).
17 Ramprashad, F.; Landolt, J.P.; Money, K.E.; Clark, D.; Laufer, J.: A morphometric study of the cochlea of the little brown bat *(Myotis lucifugus).* J. Morph. *160:* 345–358 (1979).

18 Rhode, W.S.: Observations of the vibration of the basilar membrane in squirrel monkeys using the Mössbauer technique. J. acoust. Soc. Am. *49:* 1218–1231 (1971).
19 Schuknecht, H.F.: Techniques for study of cochlear function and pathology in experimental animals. Archs Otolar. *58:* 377–397 (1953).
20 Schuknecht, H.F.: Temporal bone removal at autopsy. Preparation and uses. Archs Otolar. *87:* 33–41 (1968).
21 Spallanzani (1794), vide Dijkgraaf, S.: Spallanzani's unpublished experiments on the sensory basis of object perception in bats. Archs Otolar. *51:* 9–20 (1960).

L. Cabezudo, MD, Departamento de ORL, Centro Especial Ramon y Cajal,
Carretera de Colmenar Km. 9,100, Madrid 34 (Spain)

Adv. Oto-Rhino-Laryng., vol. 31, pp. 39–49 (Karger, Basel 1983)

Experimental Mastoidectomy with Replacement of Posterior Bony Canal Wall in Primates

Albert Hohmann

Department of Otolaryngology and Head and Neck Surgery, University of Minnesota, Minneapolis, Minn.; St. Joseph's Hospital, St. Paul, Minn., USA

Introduction

During the past two decades, otologists have obliterated the mastoid cavity and reconstructed the posterior bony canal wall in an ever-increasing number of patients, and reports of the techniques are numerous [1, 4–6, 8, 9]. However, histological evidence of the graft survival, complications, and failures are rare. In the mastoid bone, complications are usually slow to occur and the otologist who performed the original procedure may not see the failures or complications [7]. Findings from animal experiments cannot be 100% extrapolated to man; however, primate experiments done by the author [2] on mastoid cavity obliteration have shown histopathological changes (graft shrinkage, atrophy, and infection) in 2 years of postoperative survival time which were identical to those seen in humans after 5–10 years. This suggests that a survival period of 2 years in a monkey is probably equivalent to 5–10 years in man. It seems reasonable to assume that a species of primates which reaches puberty at the age of 2 or 3 years and has a life expectancy of only 20–30 years will exhibit histopathologic changes at a faster rate than *Homo sapiens*. These considerations motivated us to do another primate experiment.

Purpose of this Experiment

The purpose of this experiment was to study two questions: (1) Does temporary removal of the posterior bony canal wall with interruption of the blood supply due to the mastoidectomy and detachment of the meatal skin result in necrosis, partial absorption, or devitalization of the posterior bony wall? (2) Does the meatal skin flap become firmly and exactly reattached to the osseous segment and does the healing ear canal skin bridge the bony defects to form an adequate barrier between the external auditory canal and the mastoid cavity?

Method

In nine normal primate ears, most of the mastoid air cells were removed. The thin bony posterior canal wall segment was temporarily removed and then reimplanted after the mastoid cavity had been obliterated. The following materials were used to support the canal wall in anatomical position: Gelfoam in three ears, autogenous cancellous iliac bone, autogenous fatty tissue and blood clot in two ears each. In all animals, the mastoidectomy was performed through a postauricular incision because of the very extensive pneumatization of the mastoid. This approach also facilitated better access to the entire posterior bony canal segment. In proportion to the human anatomy, the auditory canal in monkeys is very long and narrow, approximately twice as long and one third as wide as in man. The temporalis muscle was identified and retracted upwards. Then the posterior meatal canal skin and periosteum were detached down to the posterior bony anulus. The mastoid air cells were exenterated in the usual manner using a dental drill and irrigation. The bone removal started 2–3 mm posterior to the margin of the bony canal. The bony posterior meatal wall was thinned down to leave only a 1- to 2-mm, thin, 'bony bridge' separating the mastoid cavity from the meatus. The periantral cells were removed and the medial end of the bony bridge was thinned out along the posterior aspect of the bony tympanic annulus to expose the short process and body of the incus, lateral semicircular canal, antrum threshold, and a small posterior tympanotomy slot as originally described by *Jansen* [3].

The bony posterior meatal wall was then removed by means of a superior and inferior osteotomy. The superior osteotomy extended laterally from the posterior part of the attic and ran parallel to the roof of the external ear canal. The inferior one extended laterally from the lower end of the posterior tympanotomy slot along the floor of the external ear canal. The osteotomies were started from posterior by cutting fine grooves in the bone using a small cylindrical dental fissure burr. Then just before entering the external canal, a thin and narrow rhinoplasty chisel was used to complete the osteotomy cuts. This was done to preserve cortical bone and insure more accurate fitting of the bony wall upon reinsertion. The mobile bony segment was then removed. In five ears the posterior bony canal wall was removed in one segment. In two ears, the bony wall fragmented into two pieces, and into three in another ear. In one ear, the canal bone was very brittle and shattered into more than three pieces during the osteotomies. This canal wall was reconstructed with two pieces of cortical hip bone autograft.

The removed bony wall was kept moist in a sponge soaked with normal saline solution while the remaining mastoid cells were removed. Within a period of 30 min, the bone segment or segments were reinserted into the normal anatomical position. Then the mastoid cavity was filled with either Gelfoam, blood, or the autogenous grafts. The membranous ear canal was replaced and the ear canal was packed with Gelfoam pledgets. All animals were given one million units of Bicillin.

Four ears were used for controls. In two, an identical mastoidectomy was done, but the bony posterior canal was not reinserted. The membranous canal was preserved and repositioned. In the two other ears, a simple mastoidectomy was performed; most of the air cells were removed. However, the bony canal wall was not temporarily removed and the meatal skin was not detached. The mastoid cavities were not obliterated in the four control ears. After different postoperative periods ranging from 9 to 24 months, the animals were sacrificed. The temporal bones were removed, fixed in formalin, decalcified, embedded in celloidin, and serial horizontal sections were stained in hematoxylin and eosin and examined by light microscopy.

Results

The control ears are discussed first to show the different results in mastoidectomy with preservation of the bony canal wall versus removal of the bony canal wall. In the two ears with an intact and preserved posterior bony wall, the configuration of the ear canal was normal. The bone had a typical cortical pattern and showed no signs of infection or absorption (fig. 1). High power microscopic examination demonstrated viable osteocytes. The meatal skin was normal with healthy squamous epithelium and a broad layer of periosteum with many small blood vessels (fig. 1, insert).

Removal of the bony canal produced entirely different histological findings. The membranous canal wall without bony support and nourishment atrophied to a thin membrane which either bridged the defect (fig. 2a) or retracted into the cavity (fig. 2b) creating an open mastoid cavity.

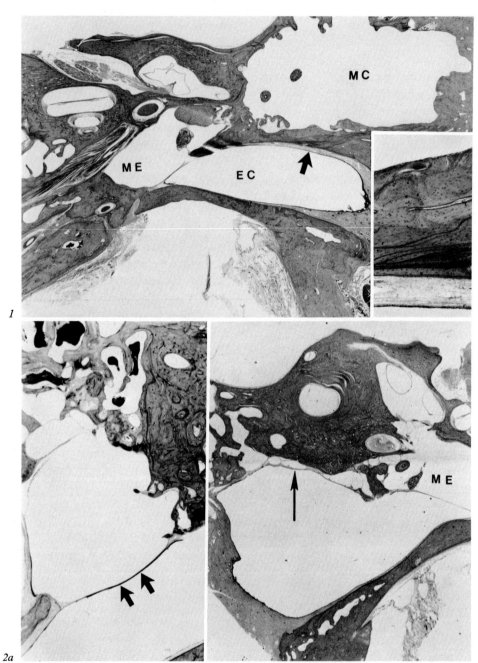

1

2a   2b

Results after Reinsertion of Bony Canal Wall

The nine mastoidectomies with reconstruction of the bony canal wall will be discussed in respect to the fate of the reimplanted bone, and degree of preservation of the ear canal skin, and complications.

The reimplanted posterior canal wall segment survived in six out of nine ears. The degree of bone survival varied and was related to the graft material used for obliteration and blood supply from the postauricular soft tissue. The two mastoids obliterated with autogenous hip bone showed the best long-term results as far as the bony canal wall and canal skin were concerned. 9 months postsurgery some cancellous bone in the center of the cavity was preserved. A major portion of the bone was replaced by fibrous tissue. No infection of the mastoid, middle or outer ear was present. The bone graft was adequate to provide permanent and exact support for placement of the posterior canal wall. The normal configuration of the bony canal wall was preserved and so was the tympanomeatal flap. Canal wall bone and mastoid cavity graft were united by bony and fibrous unions (fig. 3). On high-power examination of the canal wall, there was a marked increase of empty lacunae when advancing from lateral to medial (fig. 3, insert). In the second animal the external ear canal was reconstructed with two autogenous cortical iliac bone struts which were used as a substitute for the bony canal wall. Both segments were well preserved and united by bony union to each other. The posterior segment was also fixed to the cancellous bone graft by several bony connections (fig. 4). The medial ends of the two grafts could not be shaped precise enough to duplicate the posterior canal wall and, therefore, did not make contact with the posterior tympanic annulus. A several millimeter wide defect between the tympanic annulus and the bone grafts existed which was bridged by the tympanomeatal skin. During the epithelial migration across

---

Fig. 1. Low-power micrograph of a primate middle ear mastoid following a mastoidectomy with 'bony canal wall left up' and tympanomeatal flap not raised. EC = External canal; ME = middle ear; MC = mastoid cavity. Solid arrow marks the area of the insert with a high power view showing the canal wall epithelium with a thick periosteum and cortical canal wall with many viable osteocytes.

Fig. 2. a Low-power photomicrograph of a control ear following removal of the bony posterior canal wall. The thin canal skin stretches across the mastoid cavity defect (arrows). *b* Another control animal where the meatal skin retracted into the cavity. 1 year postmastoidectomy and canal removal the skin is loosely attached to the bone of the mastoid cavity (arrow). ME = Middle ear. Large mastoid cavity.

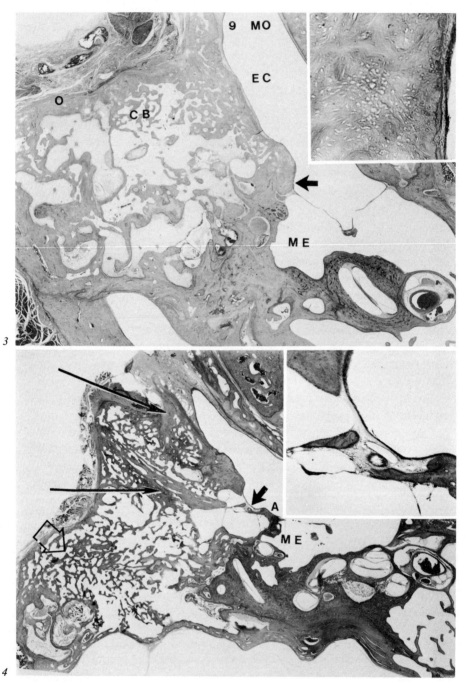

the void not covered by any type of graft material, a small epidermoid inclusion cyst developed in the tympanomeatal flap (fig. 4, insert). A second epidermoid inclusion cyst was present almost directly opposite the first one in the anterior recess between the tympanic membrane and the anterior bony canal wall. The osteocyte survival rate in the two cortical bone grafts was very similar to the one encountered in the first animal. The meatal skin covering the bony grafts was preserved. Considering the long survival time of 24 months, the appearance of both the graft and meatal skin was good. Similar findings in posterior canal wall and skin survival were seen in the two mastoids obliterated with autogenous fat grafts and in two out of three Gelfoam ears.

13 months after implantation, approximately one third of the fat graft had been absorbed, as documented by three large, circular defects or vacuolizations; two directly underneath the large lateral canal wall segment and one in the sinodural angle region. In the center portion of the cavity, some of the fat graft was surviving. The part of the graft adjoining the external soft tissue of the approach hole was being replaced by fibrous tissue and trabecular bone (fig. 5). The bony canal was reinserted in three pieces (see arrows). All three survived. There was fibrous union between the large distal and small medial segment. High power view of the cortical bone of the posterior canal wall showed marked reduction of living osteocytes. Empty lacunae were seen in the lateral canal wall segment which was not well supported by a viable graft bed, but extended across the fat graft defect (fig. 5, insert). The tympanomeatal flap bridged the canal wall segments and showed some atrophy of the epithelium.

In three mastoid cavities, Gelfoam was used for the support of the pos-

---

Fig. 3. Low-power photomicrograph of monkey ear 9 months postmastoidectomy with reinsertion of the bony canal wall. Note the cancellous bone (CB) being replaced by fibrous tissue. The periosteum of the approach hole region is lying down new bone and osteoid (O) to close the cavity laterally. The reinserted canal wall and canal skin both surviving, but at the medial aspect of the canal wall close to the tympanic annulus (arrow) empty lacunae predominate (see insert).

Fig. 4. Low-power photomicrograph of large primate mastoid demonstrating the 24 months survival of two autogenous cortical hip graft segments used for reconstruction of the posterior bony ear canal (long solid arrows). Open arrow marks the autogenous cancellous bone used for filling the mastoid cavity, some bone absorption noticeable. Note the canal wall graft did not make contact with the tympanic membrane at the tympanomeatal angle (A). The meatal skin bridged the defect and sealed the mastoid off from the external canal (solid short arrow). High-power photomicrograph insert shows this area in greater detail and also the small developing cholesteatoma and the viable bone with osteocytes.

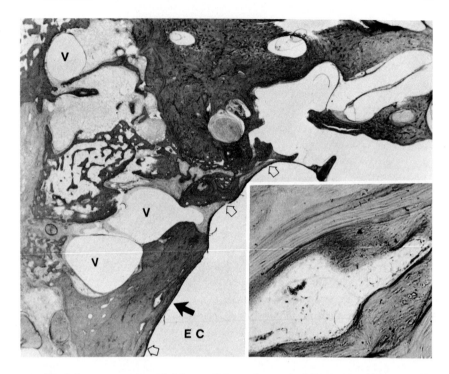

Fig.5. Low-power view of right ear (13 months survival time) mastoidectomy with autogenous fat graft for support of the reinserted posterior canal wall. Three canal wall segments (open arrows) united by bony and fibrous union. Three large areas of vacuolization (V) in the mastoid cavity and partial replacement of fat graft by new bone. High-power magnification insert of the region labeled by the solid arrows shows loss of osteocytes and a vascular channel with small blood vessel containing blood cells.

terior canal wall. In two out of three ears, the posterior canal wall and the skin survived. Gelfoam can adapt well to any type of cavity and, after air spaces are filled with blood, can give temporary support to a thin, bony canal wall. This material, however, liquifies over a period of 3–6 weeks and did not give prolonged, firm support to the canal wall. Both mastoids showed a large defect where the Gelfoam had been present and had liquified in the immediate postoperative period. In the longest surviving animal (24 months) the mastoid sections demonstrated the disadvantages of temporary obliteration with Gelfoam best (fig. 6). The Gelfoam was not replaced by ingrowth of fibrous tissue. 2 years following the surgical procedure there was still a large mastoid cavity present, which reduced in size by new bone being produced from the perios-

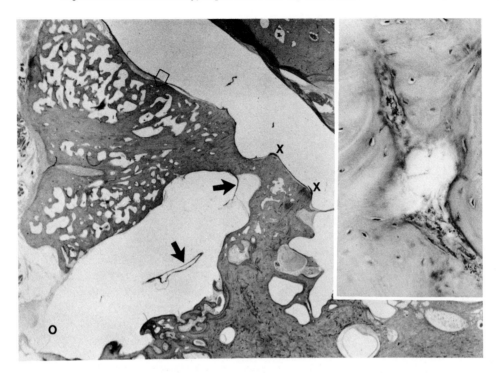

Fig. 6. Low-power photomicrograph of left primate mastoid with large cavity defect. The cavity had been filled with Gelfoam after reinsertion of the canal wall segments. Segments surviving well 2 years postoperative. Medial segment (XX) makes good contact with the ear drum at the meatal angle and also joined to the lateral segment by bony union. Posterior canal wall skin, however, extended into the mastoid cavity prior to the bony union. Solid arrows indicate ingrowth of skin and multilayered cysts in the cavity. Open arrow shows location of vascular channel with surrounding viable osteocytes shown in the insert. O = Last remaining open part of the mastoid cavity with faint staining osteoid being formed prior to new bone formation.

teum of the lateral cortex. Both mature bone and osteoid were clearly visible in the approach hole region.

On examining the canal wall implant, one gets the impression that the 'pliable' Gelfoam allowed for shifting of the two piece canal wall in the postoperative healing period. The resulting gap between the two segments was gradually bridged by bony and fibrous tissue, but only after the canal wall skin had grown into the defect and produced a cyst lined by multilayered squamous cell epithelium in the mastoid cavity (fig. 6).

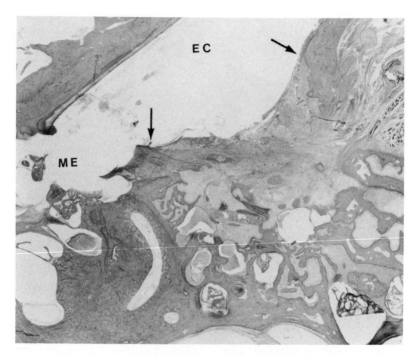

Fig. 7. Right mastoid 1 year after posterior canal wall reinsertion and filling of the cavity with a blood clot. The chronic infection and osteomyelitis has destroyd the bony canal wall to an extent outlined by solid arrows. Meatal skin missing with large ulcer and loss of tissue has produced indentation and enlargement of the posterior canal wall. Tympanic membrane missing due to artifact during sectioning of the temporal bone.

Complications

The two mastoid cavities filled with blood and one with Gelfoam were the only ones that, 12–17 months after surgery, showed extensive, chronic infection throughout the remaining mastoid bones and the middle ears. Osteomyelitis had destroyed 50–100% of the reimplanted canal wall bone. Most of the overlying canal wall skin was ulcerated. Only the outer one third of the tympanomeatal flap survived. The loss of bone and soft tissue had produced a depression in the posterior canal wall and an enlargement of the external canal (fig. 7).

Discussion

This animal experiment demonstrated that reconstruction of the bony posterior canal wall can be complicated by formation of epithelial inclusion cysts, ulceration of the canal wall skin and total or partial destruction of the bony wall. The complications were more profound when absorbable blood or Gelfoam were used. In comparing the degree of preservation of the posterior canal wall skin in this experiment with a previous one [2] where the canal wall had been taken down, there was atrophy of the canal wall skin noted, which did not exist when the meatal skin made contact with a viable tissue graft. An exact reapproximation of the bony segment at the posterior tympanic annulus was a definite problem and can also exist in humans when posterior canal wall reconstruction is attempted. The ingrowth of the canal wall skin into the mastoid cavity and recurrence of microcholesteatoma could possibly be avoided by the use of fascia or lyodura under the meatal skin as advocated by *Palva* [5]. The bone and fat grafts survived in the mastoid and kept the posterior canal wall segments in place. As could be expected, the Gelfoam and blood were liquified and absorbed too soon and were inadequate in supporting the canal wall. In three out of five ears when blood or Gelfoam was used, the canal wall was lost due to a postoperative infection.

References

1   Bennett, R.J.: The operation of tympanomastoid re-aeration. J. Lar. Otol. *95:* 1–10 (1981).
2   Hohmann, A.: Fate of autogenous grafts and processed heterogenous bone in the mastoid cavity of primates. Laryngoscope, St.Louis *79:* 1618–1646 (1969).
3   Jansen, C.: Cartilage tympanoplasty. Laryngoscope, St.Louis *73:* 1288–1301 (1963).
4   Lapidot, A.; Brandow, E.C.: A method for preserving the posterior canal wall and bridge in the surgery for cholesteatoma. Acta oto-lar. *62:* 88–92 (1966).
5   Palva, T.: Mastoid obliteration. Acta oto-lar. suppl.360, pp.152–154 (1979).
6   Palva, T.: Surgery of chronic ear without cavity. Archs Otolar. *77:* 570–580 (1963).
7   Palva, T.; Karma, P.; Kärjä, J.: Mastoid obliteration: histopathological study of three temporal bones. Archs Otolar. *101:* 271–275 (1975).
8   Portmann, M.; Hiranandani, N.: Functional results in the reconstruction of bony meatus after radical mastoid operation. Laryngoscope, St.Louis *80:* 105–110 (1970).
9   Wullstein, S.R.: Osteoplastic epitympanotomy. Trans. Am. otol. Soc. *LXII:* 148–156 (1974).

A. Hohmann, MD, Department of Otolaryngology and Head and Neck Surgery, University of Minnesota, St.Joseph's Hospital, St. Paul, MN 55102 (USA)

Adv. Oto-Rhino-Laryng., vol. 31, pp. 50–58 (Karger, Basel 1983)

The Round Window Membrane

Yasuya Nomura, Taeko Okuno (by invitation),
Isuzu Kawabata (by invitation)

Department of Otolaryngology, University of Tokyo, Tokyo, Japan

Introduction

The round window has been neglected for many years in clinical otology, except in rare cases of congenital absence of the round window or occlusion of the window due to otosclerosis. Since labyrinthine rupture has been recognized as a clinical entity, however, the round windows has drawn much attention.

There are two membranes in the round window niche area: one is the round window membrane (RWM) and the other is the round window niche membrane (RWNM). During ear surgery, it is easy to identify the RWNM in most cases. However, the RWNM can be confused with the RWM and vice versa.

The purpose of this paper is to describe the human RWM with regard to membrane rupture and to describe the RWNM from the clinical viewpoint.

Materials and Methods

150 human temporal bones were used in this series of studies. For light microscopic study, the temporal bones were removed, fixed in 10% formalin, and dissected with/without decalcification, and the RWMs were removed. The RWM was either stained or unstained and mounted in synthetic resin after dehydration in graded alcohol and xylene immersion. Conventional celloidine specimens were also made in order to observe the RWM and RWNM in the same specimens. For electron microscopic study the fresh RWM was fixed using glutaraldehyde solution and several processes were followed to make specimens for scanning as well as transmission electron microscopy.

Scala Tympani

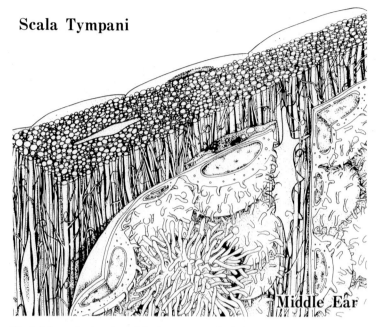

Fig. 1. Schematic drawing of the human round window membrane.

Fresh temporal bones were used for rupture experiments. For this particular purpose, a dye solution (recorder ink diluted by saline solution) was
forcefully injected into the cochlea either through the cochlear aqueduct or
through a tiny hole produced in the scala tympani of the basal turn. Ruptured
RWMs were carefully removed under the surgical microscope for light microscopic observation using the method mentioned above.

The round window niche was observed in 100 fixed temporal bones under
the surgical microscope after dissection and exposure of the round window
niche area.

Results

The RWM is round, elliptical or kidney-shaped. It is convex to the scala
tympani when observed in the removed temporal bone. It can be said that
the RWM lies deep in the RW niche and almost parallel to the floor of the
external meatus. Therefore, it is hard to observe the whole RWM by conventional anterior or posterior tympanotomy.

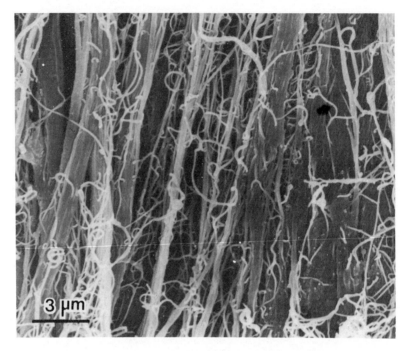

Fig. 2. Collagen fiber arrangements in the middle layer. Scanning electron micrograph.

The RWM consists of three layers: the outer layer facing the tympanic cavity, the middle layer which occupies the major part of the RWM and the inner layer facing the scala tympani (fig. 1). The outer layer can be divided into the epithelial layer and the subepithelial connective tissue layer. The latter contains the melanocytes, which can be observed during ear surgery under high power magnification. The melanocytes usually exist abundantly at the circumference of the RWM.

The collagen fibers are the main component of the middle layer, although elastic fiber and fibroblasts are also present. The collagen fibers appear to be running parallel, but actually are slightly fanning out from the crista semilunaris to the margin of the round window (fig. 2).

Fig. 3. Site and shape of experimentally induced rupture in the round window membrane. *a* There is a slit formation along the collagen fiber arrangement (arrow). *b* A high power view. A slit formation is obvious. However, because of fiber arrangements, the shape is not simple, but an intricate, slit.

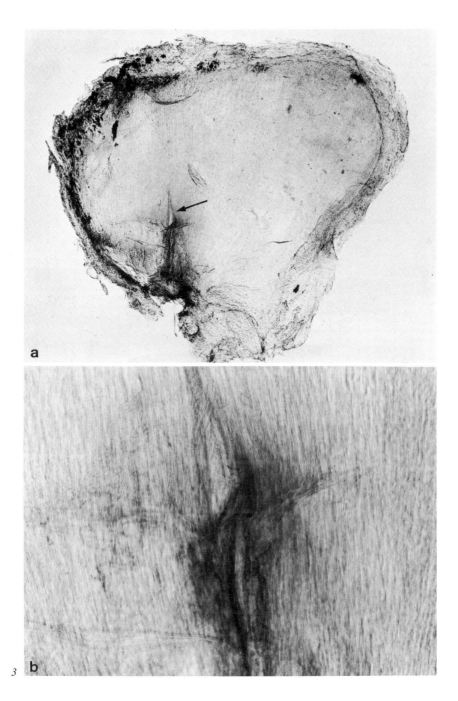

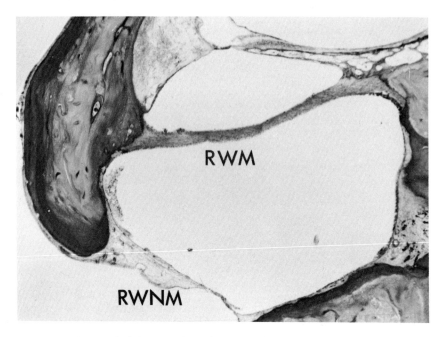

Fig.4. The round window membrane (RWM) and the round window niche membrane (RWNM).

The dye solution was injected into the cochlea in order to observe the shape and site of the rupture in the RWM. 25 temporal bones were used in this 'via cochlear aqueduct' study. No dye solution was found in 13 temporal bones. Seven stained without leaking of the solution in both windows. Three showed rupture in the oval window and two in the RWM. In the second 'direct injection' study in ten temporal bones, seven had oval window leaking, two showed RWM rupture and one stained without leaking

Four ruptured RWMs were removed for observation from the both experiments. All RWMs showed a slit along the direction of the collagen fibers and located in the central zone of the RWM, i.e. somewhere between the crista semilunaris and the margin of the window (fig. 3a, b).

Next, the round window niche was observed under the operating microscope in order to observe the presence or absence of the RWNM and the shape of perforation in the RWNM, if present. Of the 100 adult temporal bones analyzed, 30 had clear round window niches, i.e. there was no RWNM. This is the open type. The remaining 70 temporal bones showed varying types of the RWNM.

The RWNMs were classified into three types. The closed type RWNM covers the round window niche without having any perforation. The perforated type was so named because the RWNM had one or two perforations. The reticulated type had many perforations and could also be called network type. Among 100 temporal bones studied, we found open type in 30% as mentioned above, closed type in 13%, perforated type in 54%, and reticulated type in 3% (fig. 4, 5).

There are several ways of differentiating the RWM from the RWNM. (1) The RWM is located deep in the niche. This prevents direct observation. The anterior part of the RWM may be observed without curetting the bony rim. The RWNM is located at the orifice of the niche. (2) The RWM is mobile with the stapes; it moves when the stapes or long process of the incus is gently touched, whereas the RWNM does not. (3) The presence of the melanocytes serves as a good landmark of the RWM. The RWNM usually has none, or many fewer melanocytes if present. (4) If a membrane has a round perforation, it is the RWNM. The ruptured RWM usually has a slit, rather than a perforation. The exact site and shape of the rupture is hard to identify under the operating microscope. If the fluid seeps through the RWM when the stapes is gently pressed, there must be a rupture in the RWM.

Discussion

Many factors may be involved in the perforation or rupture of the RWM. Resistance of the RWM depends on not only thickness of the membrane, but amount, distribution and arrangement of collagen fibers. The direction of the collagen fibers in the middle layer is such that all fibers lead from the crista semilunaris to the margin of the round window. They fan out slightly, and several layers are superimposed. The shape of the rupture observed in figure 3b proved this. Therefore, the slit observed in the rupture experiments is not a simple, but an intricate, one.

Rupture of the RWM may not bring about severe hearing loss. Experimental studies revealed that an artificial tear in the RWM of the guinea pig did not result in much electrical change in cochlear microphonics and action potentials [2]. Simultaneous rupture of the membranous labyrinth is very likely to occur. The variety of symptoms observed in patients with round window membrane rupture cannot be explained simply. In this regard, the name 'round window membrane rupture syndrome, is more appropriate than RWM rupture [8] Animal experiments have been conducted along this line to find

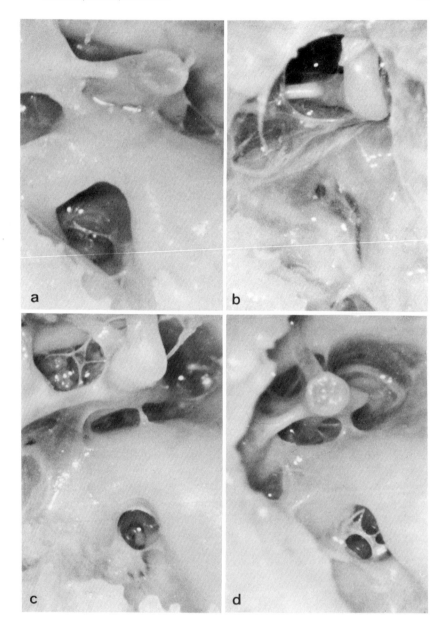

Fig.5. Types of round window niche membrane: *a* open type; *b* closed type; *c* perforated type; *d* reticulated type.

out whether or not there is a 'locus minoris resistentiae' in the membranous labyrinth.

Goodhill [3] proposed 'the explosive and implosive routes' theory to explain the mechanism of window rupture. The present experiment tried to make a simulation of the explosive route in the human temporal bones using a dye solution injected into the cochlear aqueduct. In 13 stained temporal bones the intracochlear pressure was insufficient to produce rupture in the windows. The literature indicates that the oval window is more susceptible than the round window in clinical cases [6]. The present experiments also showed a tendency for fistula to occur in the oval window. Of the oval window ruptures observed in seven temporal bones with 'direct injection', three showed in the upper middle annular ligament, three in the lower middle and one in the posterior part.

The round window area shows a wide range of anatomical variations. Some patients show a clear round window niche. The RWM can be partly observed at its anterior part by elevating the tympanomeatal flap. On the other hand, the round window niche itself cannot be seen in some cases, because of the overhanging of the anterior and superior wall of the round window niche. In the latter case, presence or absence of the RWNM cannot be confirmed without some sort of manipulation.

Several papers have described the presence of round perforation in the round window membrane in patients [1,7]. From the present study and our own observation of surgical cases, the RWM showed its shape of rupture as a slit. There may be a possibility of confusing a perforated type RWNM with the RWM in rupture patients.

It is mandatory to perform hypotympanotomy and remove the overhang of the niche in order to observe the entire RWM. Gentle touching of the stapes produces seeping of the perilymph. This is a useful diagnostic procedure even· if no rupture is observed.

As to the origin of the RWNM, it may be a remnant of the posterior sac of the embryological tissue [5]. The closed type RWNM quite frequently has a similar membrane over the stapes too.

The presence of melanocytes in the RWM has been known for many years. *Wolff* [9] also mentioned the presence of melanocytes. We demonstrated the pigment to be melanin by using two histochemical methods. The presence of melanin in a cell does not conclusively show the cell to be a melanocyte. However, the presence of premelanosomes and melanosomes was confirmed under the electron microscope. The melanocytes are observed not only in the RWM, but also in the mucous membrane of the round window niche, particularly at the base of the niche. *Gussen* [4] places special emphasis on the function of the melanocyte and webby tissues in the round window niche in relation to fluid circulation. The presence of the melanocytes might show some embryological relationship between the neural tissue and this particular middle ear tissue, judging from the fact that the round window niche area is closely related to the tympanomeningeal fissure.

References

1   Allam, A.F.: Rupture der Membran des runden Fensters. Lar. Rhinol. *55:* 544–548 (1976).
2   Fukaya, T.; Nomura, Y.: Experimental round window membrane rupture. Electrophysiological consequences in guinea pig. Audiology *24:* 152–155 (1981).
3   Goodhill, V.: Sudden deafness and round window rupture. Laryngoscope, St. Louis *81:* 1462–1474 (1971).
4   Gussen, R.: Round window niche melanocytes and webby tissue. Archs Otolar. *104:* 662–668 (1978).
5   Proctor, B.: The development of the middle ear spaces and their surgical significance. J. Lar. Otol. *78:* 631–648 (1964).
6   Thompson, J.N.; Kohut, R.I.: Perilymph fistulae: variability of symptoms and results of surgery, Otolaryngol. Head Neck Surg. *87:* 898–903 (1979).
7   Tonkin, J.P.; Fagan, P.: Rupture of the round window membrane. J. Lar. Otol. *89:* 733–756 (1975).
8   Weisskopf, A.; Murphy, J.T.; Merzenich, M.M.: Genesis of the round window rupture syndrome: some experimental observations. Laryngoscope, St. Louis *88:* 389–397 (1978).
9   Wolff, D.: Melanin in the inner ear. Archs Otolar. *14:* 195–211 (1931).

Y. Nomura, MD, Department of Otolaryngology, University of Tokyo, Tokyo (Japan)

Adv. Oto-Rhino-Laryng., vol. 31, pp. 59–71 (Karger, Basel 1983)

The Ventral Cochlear Nucleus in Human Brains

E. Richter

ENT-Department, AKH-Linz, Linz, Austria

Introduction

Different types of neurons have been described in the ventral cochlear nucleus (VCN) of the cat in a study mainly based on observations in Nissl stained material by *Osen* [26]. Different cell types were shown to occupy specific regions of the VCN. Large spherical cells for example were found in the anterior region globular cells in the interstitial cochlear nucleus (ICN), and multipolar cells posterior to the cochlear nerve trunk.

Homologous populations of neurons were observed in the VCN of other mammalian species [10–12, 20–22, 39]. Spherical and globular cells were shown to receive large somatic endings – the end-bulbs of Held – from ascending branches of the cochlear nerve, while the somata of multipolar and octopus cells were contacted by few, small terminals from the descending branches [4, 7, 20–22, 27, 28, 30]. The large end-bulbs made multiple synaptic contacts on cells in the VCN of the cat [18, 36]. Signal transmission through such large synaptic elements is probably one-for-one [3] and the field potentials which they generate may be important determinants of the shape of brain stem-evoked responses.

Under the light microscope the cytoarchitecture of the human cochlear nucleus was found similar to that in lower mammalian species. Most principal neuronal types could be identified, but their distribution differed and subdivisions were found less well-defined in man [2, 5, 6, 8, 9, 15, 23, 25]. It remained unclear, however, whether spherical and globular cells receive large end-bulbs of Held from ascending cochlear branches as in other mammalian species. For the interpretation of the results from animal studies and for a better understanding of the function of human auditory centers, the answer to this question is of some importance.

Methods

The distribution of cell types and large endings in the human VCN was first studied in conventional Nissl- and protargol-stained material from the collection of the Eaton Peabody Laboratory at the Massachusetts Eye and Ear Infirmary. From serial, 20-μm sections of one of these brain stems in a transverse plane, photomicrographs were taken and the cochlear nucleus traced in every fourth section. Registration holes perpendicular to the sectioning plane permitted a reconstruction of auditory centers in a horizontal and sagittal plane. The location and dimensions of the human cochlear nucleus (CN) in relation to the bony labyrinth and brain stem can be seen in these projections (fig. 1, 2).

Three adult brains were prepared for electron microscopy. The brains were removed and perfused through the carotids at 2, 3 or 5 h after death with 500 ml normal saline followed by 30 min of a solution containing 1.5% glutaraldehyde and 2.5% paraformaldehyde in a 0.1 M phosphate-buffered solution with 0.005% $CaCl_2$ at a pH pf 7.2 [14]. After removal of the hemispheres and most of the cerebellum, the CN was sliced on one side by several sections in a transverse plane at a distance of about 2 mm from each other before immersion in the above fixative overnight. Several 2-mm blocks were then removed and the remainder of the brain stem saved for other purposes. The blocks were bathed in a 1% solution of osmium tetroxide in the above buffer, before dehydration and embedding in Spurr's resin. Thick sections at 2 μm were made in a transverse plane and stained with toluidine blue for light microscopic identification of the VCN. Structures serving as landmarks were the cochlear nerve trunk, containing only few perikarya, the cerebellum and the ventral acoustic stria, which can be seen in representative atlas sections (fig. 3, 4). After orientation of large thick sections under the light microscope, the blocks were trimmed, sectioned for electron microscopy and contrasted with uranyl acetate and lead citrate [31].

Results

In the cat, different locations were described for small and large spherical cells [26]. The diameters of cells in the anterior VCN from human brains showed a unimodal distribution [2]. In regions anterior to the cochlear nerve root, spherical cells predominated with diameters varying between 25 and 35 μm. Spherical cells are round or ovoid in shape and contain coarse Nissl

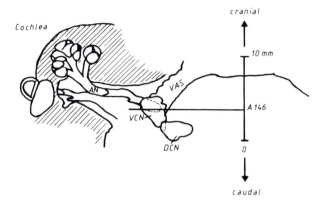

Fig. 1. Horizontal projection of the cochlear nucleus. The location of the atlas section 146 is indicated. Schematic drawing of the extraglial portion of the acoustic nerve in the inner ear canal derives from a midmodiolar horizontal temporal bone (shaded area) section. The arrows pointing cranial and caudal represent the midline of the brain stem. A146 = Atlas section 146 (at 20 μm); VAS = ventral acoustic stria; VCN = ventral cochlear nucleus; DCN = dorsal cochlear nucleus; AN = Acoustic nerve.

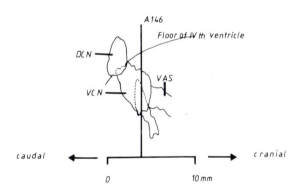

Fig. 2. Sagittal projection of the cochlear nucleus. A dark line indicates the location of atlas section 146. For abbreviations, see figure 1.

substance arranged concentrically around a central nucleus. Globular cells are found adjacent to the cochlear nerve trunk intermingled with spherical and multipolar cells. Characteristically they contain an eccentric nucleus, fine dispersed Nissl substance and have about the same size range as spherical cells. Multipolar cells have a polygonal shape, and contain coarse Nissl substance and a central nucleus. They show about the same range in diameters as spheri-

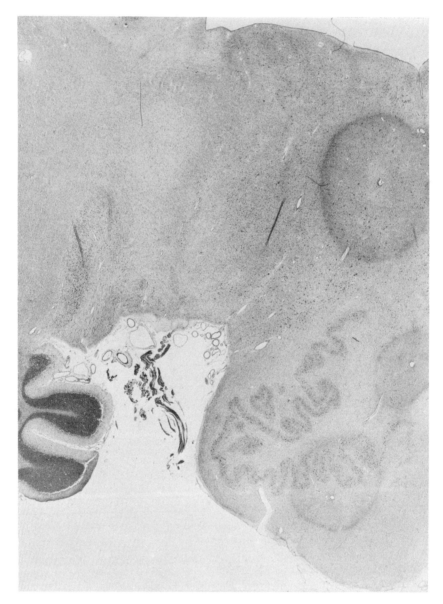

Fig. 3. Transverse brain stem section (atlas section 146). (C = Cerebellum; IO = inferior olive; ICN = interstitial cochlear nucleus. Further abbreviations as in figure 1.

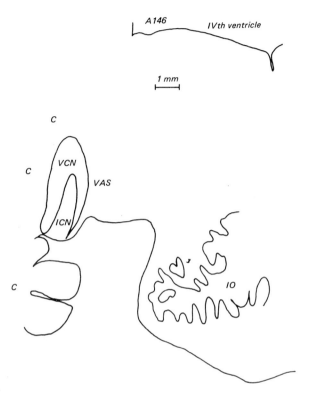

Fig. 4. Tracing of the outlines of the structures seen in figure 3.

cal and globular cells and populate the largest area of the VCN posterior to
the cochlear nerve trunk (fig. 5a–c). One of the most prominent features of
the CN in cats are the large end-bulbs of Held seen on the spherical and globular cells. The appearance of globular cells in protargol-stained human material is illustrated in figure 6. In such material, the presence of end-bulbs of
Held is much more ambiguous than in similar preparations from the cat. However, some possible examples are indicated by the arrows in figure 6.

Based on morphological characteristics such as their shape, proximal
dentrites, and nuclear positions, spherical, globular and multipolar cells are
identifiable under the electron microscope at low magnification. Synaptic endings can be demonstrated using higher magnifications. Large endings contacting a great portion of perikarya in the AVCN, like an egg-cup may cover an
egg, were called end-bulbs of Held. These characteristic cups, however, were
found to be present in young cats only and to change later during maturation.

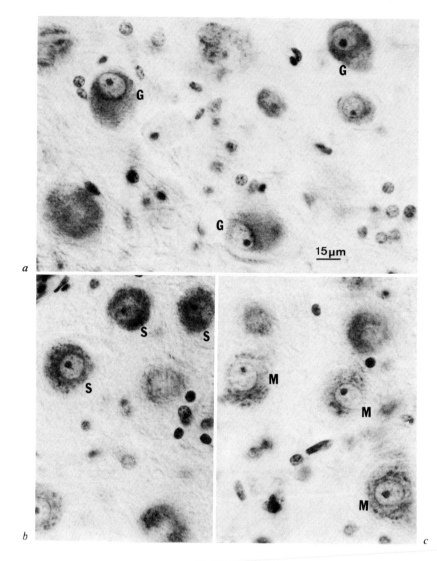

Fig.5. a–c Nissl-stained cells of the ventral cochlear nucleus (VCN). Globular (G), spherical (S) and multipolar (M) cells are indicated.

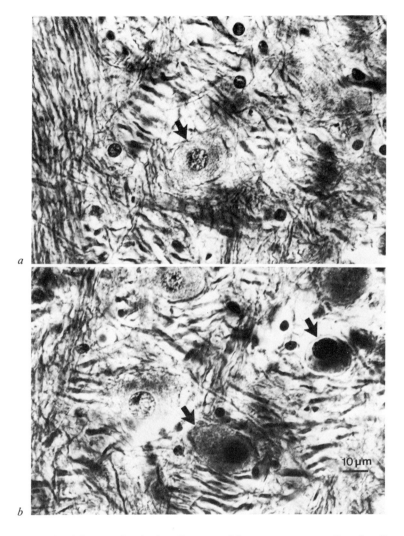

Fig.6. a,b Protargol stained perikarya receiving structures suggestive of endings are indicated by arrows.

In the adult cat the axons branch on the somata to form basket-like arborizations covering the cell bodies [22, 33]. In a given thin section from globular or spherical cells, endings cover almost the entire soma (fig. 7, 8). The profile of such an ending may appear cup-like if the sectioning plane corresponds with the axis of the terminal. Semiserial sections, however, reveal a diameter

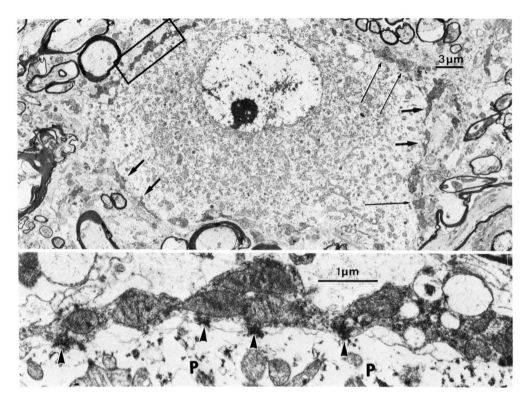

3 µm

1µm

P   P

Fig. 7. A low magnification electron micrograph of a cell from the AVCN with an eccentric nucleus. The surface is covered by long, thin, almost black, endings (thick arrows) and some lighter boutons (thin arrows). Inset shows a long and thin ending containing many mitochondria and synaptic vesicles at higher magnification. On the perikaryal (P) side triangles point to synaptic membrane specializations.

of up to 2 µm for these endings and repeated branching. In some instances a myelinated axon can be identified, losing the myelin sheath before ending on the perikaryon with multiple synaptic sites (fig. 8). *Lenn and Reese* [18] found both end-bulbs and boutons on cells in the AVCN of several mammalian species. Both types of endings formed synapses containing vesicles in the presynaptic terminal and asymetrical pre- and postsynaptic membrane densities. Based on measurements of vesicle diameters the above authors could differentiate end-bulbs from boutons. The human brains were not well enough preserved to measure vesicles or the thickness of membrane densities. The end-bulbs, however, always appeared dark (fig. 7, 8) in the present material, while boutons, which had no connection to end-bulbs, if traced in semiserial

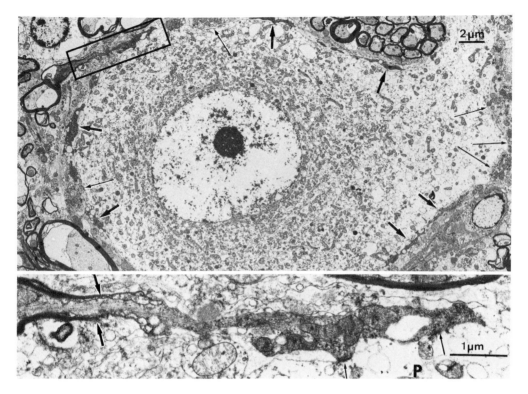

Fig. 8. Electron micrograph of a cell from the AVCN with a central nucleus. Long, thin and almost black endings (thick arrows) and lighter boutons (thin arrows) cover the surface. Inset shows an axon, loosing its myelin sheath (thick arrows) before contacting (thin arrows) the perikaryon (P) at higher magnification.

sections, had a lighter appearance (fig. 9). Finally, multipolar cells predominating in the region posterior to the cochlear nerve trunk were almost devoid of somatic endings (fig. 10).

Discussion

The distribution of principal cell types in the VCN of human brains in this study is not much different from that found in earlier ones by other authors based on Nissl stains [16, 17, 23]. In protargol-stained sections from human brains a group of 'spheroid' neurons was observed in the anterior VCN, which received large end-bulbs [5, 6]. It has been suggested that such

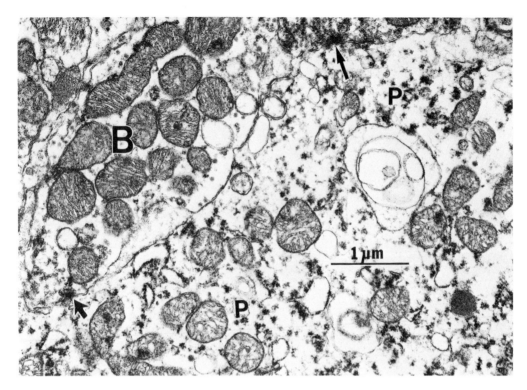

Fig.9. Electron micrograph showing a light bouton (B) contacting a cell (P) (short arrow). The synaptic contact of a darker thin and long ending is indicated by a thinner arrow.

end-bulbs may not be restricted to spherical cells since they were also observed caudal to the nerve root [23]. In this study structures suggestive of end-bulbs could be found on many perikarya in protargol-treated sections from the VCN. A really convincing example could not be demonstrated. Therefore, specimens from the area in question were viewed under the electron microscope. Morphological criteria used for classification of neurons in the Nissl-treated material, such as configuration of the Nissl substance, shape, dentritic pattern and nuclear position are as well applicable in thin sections. As shown in the cat [36] multipolar cells received few small somatic endings, while globular and spherical cells were contacted by both end-bulbs with multiple synaptic sites and boutons. The bulbs were similar in shape to calyces found on principal cells of the human MNTB [32] but much smaller and derived from much thinner axons than the latter.

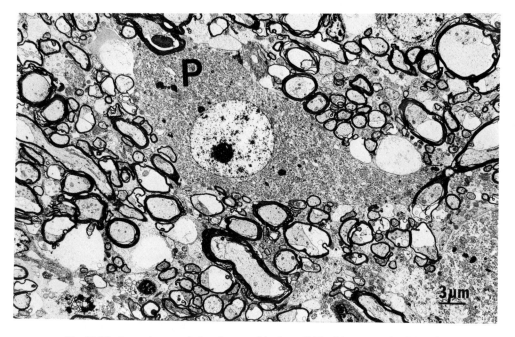

Fig. 10. Electron micrograph showing a multipolar cell (P) with two proximal dentrites, almost devoid of somatic endings.

Adult brains were available, only, for the present study. The shape of end-bulbs was not cup-like, but similar to their form in the adult cat [20,33]. In the cat globular cells were shown to give rise to thick axons that cross in the trapezoid body to end on principal cells of the MNTB on the contralateral side with calyces of Held [13,24,35,36]. Since there is good evidence for the presence of an MNTB in human brains [32] there could be similar connections in human brains. Axons of large spherical cells in the anterior VCN of the cat provide bilateral input to the medial superior alive (MSO) [27,28,37,38] by long terminals with many synaptic contacts on dentrites and somata of MSO neurons [19,34]. The axons of multipolar cells were shown to project mainly to the inferior colliculus [1,29]. Thus, in man the population of neurons occupying the greatest part of the VCN with few small somatic endings and thin axons – the multipolar cells – project mainly to higher levels of the brain stem, while a relatively small population as compared to the cat, with large synaptic endings and thick axons – the globular and spherical cells – project to the superior olivary nuclei, as already suggested by *Moore and Osen* [23].

The demonstration that the human VCN contains cell types with a similar distribution and synaptic organization as reported for lower mammals [10–12, 26] has important consequences, and it could well be that results from electrophysiological studies on the CN in animals may be applicable in man.

References

1   Adams, S.C.: Organization of the margins of the auteroventral cochlear nucleus. Anat. Rec. *187:* 520 (1977).
2   Bacsik, R.D.; Strominger, N.L.: The cytoarchitecture of the human anteroventral cochlear nucleus. J. comp. Neurol. *147:* 281–290 (1973).
3   Bourk, T.R.: Electrical response of neural units in the anteroventral cochlear nucleus of the cat; Ph D diss., Cambridge (1976).
4   Brawer, J.R.; Morest, D.K.: Relations between auditory nerve endings and cell types in the cat's anteroventral cochlear nucleus seen with the Golgi method and Nomarski optics. J. comp. Neurol. *160:* 491–506 (1975).
5   Dublin, W.B.: Cytoarchitecture of the cochlear nuclei. Report of an illustrative case of erythroblastosis. Archs Otolar. *100:* 355–359 (1974).
6   Dublin, W.B.: Fundamentals of sensorineural auditory pathology (Thomas, Springfield 1976).
7   Feldman, M.L.; Harrison, S.M.: The projection of the acoustic nerve to the ventral cochlear nucleus of the rat. A Golgi stuy. J. comp. Neurol. *137:* 267–294 (1969).
8   Fuse, G.: Das Ganglion ventrale und das Tuberculum acusticum bei einigen Säugern und beim Menschen. Arb. Hirnanat. Inst. Zürich *7:* 1–210 (1913).
9   Hall, J.G.: The cochlea and the cochlear nuclei in neonatal asphyxia. Acta oto-lar. suppl. 194 (1965).
10 Harrison, J.M.; Irving, R.: The anterior ventral cochlear nucleus. J. comp. Neurol. *124:* 15–42 (1965).
11 Harrison, J.M.; Irving, R.: Ascending connection of the anterior ventral cochlear nucleus in the rat. J. comp. Neurol. *126:* 51–64 (1966).
12 Harrison, J.M.; Irving, R.: The organization of the posterior ventral cochlear nucleus in the rat. J. comp. Neurol. *126:* 391–403 (1966).
13 Held, H.: Die zentrale Hörleitung. Arch. Anat. Physiol. Anat. Abt. *17:* 201–248 (1893).
14 Karnovsky, M.J.: A formaldehyde-glutaraldehyde fixative of high osmolarity for use in electron microscopy. J. Cell Biol. *17:* 137A (1965).
15 Konigsmark, B.W.: Neuronal population of the ventral cochlear nucleus in man. Anat. Rec. *163:* 213 (1969).
16 Konigsmark, B.W.: Cellular organization of the cochlear nuclei in man. J. Neuropath. exp. Neurol. *32:* 153–154 (1973).
17 Konigsmark, B.W.: Neuroanatomy of the auditory system. Archs Otolar. *98:* 397–413 (1973).
18 Lenn, N.J.; Reese, T.S.: The fine structure of nerve endings in the nucleus of the trapezoid body and the ventral cochlear nucleus. Am. J. Anat. *118:* 375–389 (1966).
19 Lindsey, B.G.: Fine structure and distribution of axon terminals from the cochlear nucleus on neurons in the medial superior olivary nucleus of the cat. J. comp. Neurol. *160:* 81–104 (1975).

20   Lorente de Nó, R.: Anatomy of the eight nerve. The central projection of the nerve endings of the internal ear. Laryngoscope, St. Louis *43:* 1–38 (1933).

21   Lorente de Nó, R.: Anatomy of the eighth nerve. III. General plan of structure of the primary cochlear nuclei. Laryngoscope, St. Louis *43:* 327–350 (1933).

22   Lorente de Nó, R.: Some unresolved problems concerning the cochlear nerve. Ann. Otol. Rhinol. Lar. *34:* suppl., pp. 1–28 (1976).

23   Moore, J.K.; Osen, K.K.: The cochlear nuclei in man. Am. J. Anat. *154:* 393–419 (1979).

24   Morest, D.K.: The collateral system of the medical nucleus of the trapezoid body of the cat, its neuronal architecture and relation to the olivo-cochlear bundle. Brain Res. *9:* 288–311 (1968).

25   Moskowitz, N.: Comperative aspects of some features of the central auditory system of primates. Ann. N.Y. Acad. Sci. *167:* 357–369 (1969).

26   Osen, K.K.: Cytoarchitecture of the cochlear nuclei in the cat. J. comp. Neurol. *136:* 453–484 (1969).

27   Osen, K.K.: Afferent and efferent connections of three well-defined cell types of the cat cochlear nuclei; in Anderson, Jansen. Excitatory synaptic mechanismus, pp. 295–300 (Universitetsforlaget, Oslo 1970).

28   Osen, K.K.: Course and termination of the primary afferents in the cochlear nuclei of the cat. An experimental anatomical study. Arch. ital. Biol. *108:* 21–51 (1970).

29   Osen, K.K.: Projection of the cochlear nuclei on the inferior colliculus in the cat. J. comp. Neurol. *144:* 355–372 (1972).

30   Ramón y Cajal, S.: Histologie du système nerveux de l'homme et de vertébrés (1972 2nd ed.), vol. I, pp. 774–838 (Instituto Ramón y Cajal, Madrid 1909).

31   Reynolds, E.S.: The use of lead citrate at high pH as an electron-opaque stain in electron microscopy. J. Cell Biol. *17:* 208–212 (1963).

32   Richter, E.; Norris, B.E.; Levine, R.A.; Kiang, N.Y.S.: The medial nucleus of the trapezoid body in human brains. Submitted to Am. J. Anat.

33   Fekete, D.; Ryugo, D.K.: Development of an auditory nerve terminal: The endbulb of Held (Abstract) Soc. for Neurosci. *6:* 818 (1980).

34   Schwartz, I.R.: Axonal endings in the cat medial superior olive: coated vesicles and intercellular substance. Brain Res. *46:* 187–202 (1972).

35   Stolter, W.A.: An experimental study of the cells and connection of the superior olivary complex of the cat. J. comp. Neurol. *98:* 401–431 (1953).

36   Tolbert, L.P.: Synaptic organization in the anteroventral cochlear nucleus of the cat: a light and electron microscope study; Ph D diss., Cambridge, pp. 1–231 (1978).

37   Van Noort, J.: The structure and connection of the inferior colliculus. An investigation of the lower auditory system (Van Gorcum, Assen 1969).

38   Warr, W.B.: Fiber degeneration following lesions in the anterior ventral cochlear nucleus of the cat. Expl. Neurol. *14:* 453–474 (1966).

39   Webster, B.B.; Ackermann, R.F.; Longa, G.C.: Central auditory system of the kangeroo rat *Dipodomys merriami.* J. comp. Neurol. *133:* 477–494 (1968).

E. Richter, MD, ENT-Department, AKH-Linz, Krankenhausstrasse 9,
A–4020 Linz (Austria)

Adv. Oto-Rhino-Laryng., vol. 31, pp. 72–84 (Karger, Basel 1983)

The Fine Surface View of the Adult Human Eustachian Tube in Normal and Pathological Conditions

F. Hiraide[1]

Department of Otolaryngology, National Defense Medical College, Saitama, Japan

Introduction

The Eustachian tube is one of the most complicated structures in the human body. Its anatomy is certainly very complicated, its physiology is still incompletely understood, and its pathological states are infrequently examined. The mucous membrane of the tube is known to be a direct continuation of the nasopharyngeal mucosa. Its epithelium consists of pseudostratified ciliated cells, non-ciliated columnar cells, goblet cells and underlying basal cells resting on a thin basement membrane. The author has previously investigated the density and the distribution of cilia-bearing cells in the normal human adult eustachian tube by means of scanning electron microscopy (SEM) [4]. In this study the fine surface morphology of the pathological Eustachian tube is compared with that of the normal Eustachian tube.

Material and Method

19 Eustachian tubes were used in this study. Twelve normal tubes were removed from the temporal bones of 7 patients (4 cases of gastric cancer, 2 cases of laryngeal cancer and 1 case of Hodgkin's disease). These subjects had neither clinical signs of middle ear diseases nor histories of otitis media. Seven pathological tubes were obtained from the temporal bones of 5 patients with maxillary and thyroid cancer. These patients had serous middle ear

[1] The author wishes to express his sincere thanks to Miss *Patricia Schachern* for correcting the English, and Miss *Kim Flanders* for typing this manuscript.

Fig.1. In the normal Eustachian tube, the majority of the epithelial cells on the floor and lower parts of the tubal walls are ciliated cells. Goblet cells are noted among ciliated cells. × 133.

effusions at the time of their death. The temporal bones were obtained from adult patients with ages ranging from 45 to 86 years. Within 10–120 min postmortem, 10% formalin solution was instilled through the tympanic membrane into the middle ear cavity. 3–11 h later the temporal bones including the whole length of the Eustachian tube were removed and transferred to a 10% formalin solution. Temporal bones were stored at room temperature for a period ranging from 2 days to 3 weeks. The entire Eustachian tubes were then dissected from the temporal bones under the operating microscope. The Eustachian tubes were divided into several pieces and fixed again in either 10% formalin or 1% osmic acid for approximately 24 h. Each specimen was dehydrated in graded series of alcohol, placed in isoamyl acetate for approximately 30 min, and critical point-dried using carbon dioxide [13]. The surface of the mucous membranes of the dissected Eustachian tubes was coated in carbon and gold (200–300 Å) in a vacuum evaporator on a rotating stage and examined with the SEM (JSM-T 20) at magnifications of 35–10,000 times.

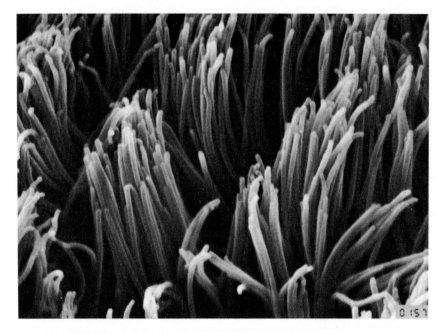

Fig. 2. High magnification of figure 1. The cilia are dense, long, and uniform in length. Their average length is approximately 8 μm. × 3300.

Results

Normal Eustachian Tube. The density and the distribution pattern of ciliated cells were determined by observing the cilia on the luminal surface of the tubal mucosa of the entire length of the Eustachian tube. The majority of epithelial cells covering the lumen of the medial one third of the tube from the pharyngeal orifice were ciliated cells (fig. 1, 2). In the middle one third of the tube, toward the roof of the tubal lumen the number of ciliated cells gradually diminished. However, on the floor and lower parts of the tubal walls, epithelial cells were mostly of the ciliated type. In the lateral one third of the tube from the isthmus to the tympanic orifice, most of the epithelial cells, covering the upper one third of the tubal walls, were non-ciliated cells (fig. 3). These non-ciliated cells appeared from their characteristic surface structures as squamous and cuboidal epithelial cells. However, the epithelial cells on the floor and lower parts of this portion frequently contained cilia. The distribution pattern of ciliated cells in the entire Eustachian tube is shown

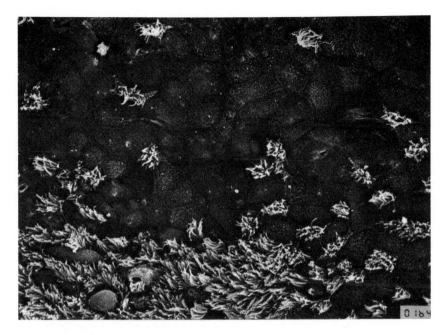

Fig. 3. The epithelial cells covering the roof and the upper parts of the tubal walls are non-ciliated cells. They are deemed as squamous and cuboidal cells from the characteristic surface structures. × 67.

in figure 4. Most ciliated cells possessed about 100–200 cilia. These cilia were dense, long, and almost uniform in length. The average length of the cilia was approximately 8 μm. There was no difference in the density of the ciliated cells between the anterior and posterior walls. Short cilia-bearing cells were infrequently noted among the ciliated cells. The length and number of these cilia were variable, but in general appeared to be almost half of the length and number of ordinary ciliated cells. Among ciliated cells, goblet cells in various secretory phases were observed. The ratio of ciliated cells to goblet cells was approximately 6:1. Goblet cells were more frequently found in the ciliated cell area than in the non-ciliated cell area. The surface shape and size of most ciliated cells were the same throughout the tubal lumen. However, the surface structure of the non-ciliated cells, including the goblet cells, varied in shape and size. Goblet cells, in particular, varied in size, the majority being 5–10 μm. Small goblet cells 2–3 μm, and rather large goblet cells measuring 15 μm in diameter were observed. Goblet cells larger than 20 μm, even in

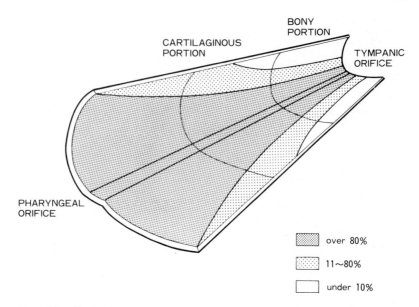

BONY
PORTION

CARTILAGINOUS
PORTION

TYMPANIC
ORIFICE

PHARYNGEAL
ORIFICE

over 80%

11~80%

under 10%

Fig. 4. The distribution pattern of the ciliated cells in the entire Eustachian tube is shown as a scheme. High density of the ciliated cells is noted on the floor and lower parts of the tubal walls.

active secretory function, were seldom seen. Ductal openings of the mucous glands were sometimes found among the ciliated epithelial cells especially in the cartilaginous portion of the Eustachian tubes.

Pathological Eustachian Tube. Although the superficial mucus resting on the top of the cilia concealed in part the underlying structures, the fine surface of the tubal mucosa could be investigated. The density and the distribution pattern of the ciliated cells in the tubal mucosa of serous middle ear effusions were basically identical with those of normal mucosa. However, the surface view of the tubal mucosa was extremely altered in the pathological state.

In general, the surface of the ciliated epithelial mucosa became uneven and irregular because of the tremendously active secretory function of the goblet cells (fig. 5). A large number of secreting goblet cells were observed in the ciliated area of the Eustachian tube. These cells appeared 3–5 times larger than ordinary cells, due to bulging of the cell surface. Cells measuring more than 20 μm were frequently observed. Goblet cell bulging caused compression and deformity of the ciliated cells. Approximately 70% of the cilia showed

Fig.5. In the inflammed Eustachian tube, the most striking findings are high activities of mucus secretion by goblet cells. The ciliated cells are compressed by bulging goblet cells. ×2330.

an irregularity in the orientation of the cilia. Ciliated epithelial cells had a tendency to loose cilia. If present, the number of the ciliary processes per cell was reduced. Their length varied from short stubby structures to fully developed cilia (fig. 6). A considerable number of knob formations was found on the ciliary surface of some ciliated cells. These structures were mainly located at the tip and upper part of the cilia (fig. 7). Compound cilia with a diameter of approximately 1 μm were occasionally detected in the inflammed tubal mucosa (fig. 8). These were probably produced by the fusion of several cilia.

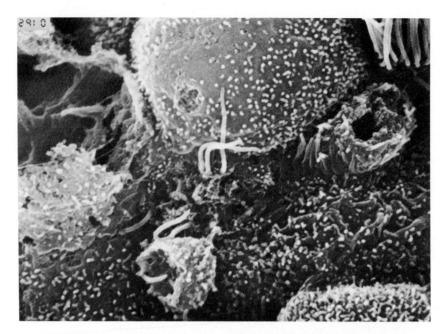

Fig.6. Decreased number of the cilia is noted in some ciliated cells. Most cilia are missing. × 3300.

Short cilia were found more frequently in the pathological tube than in the normal tube. Elongated, thin cilia were also observed but infrequently (fig. 9, 10).

These abnormal cilia were noted in approximately 10–20% of all cilia observed. The surface shape and size of squamous and cuboidal cells on the roof of the bony portion of the tube were remarkably irregular when compared with those of the normal tube.

Microvilli of most ciliated cells were normal in number and structure. However, a marked elongation of the microvilli was noted in some cells. Others showed club-like or bud-like formations of varying size on the tips or sides of the microvilli. Short microvilli were usually noted in the ciliated cells with reduced numbers of cilia. The number and shape of the microvilli of the non-ciliated cells were variable even in the normal tubal mucosa. Therefore, a comparison could not be carried out in both normal and pathological conditions. There was no granulation in the lumen of any of the Eustachian tubes examined.

Fig. 7. A considerable number of knob formations (arrow) are found on the ciliary surface of some ciliated cells. × 5000.

Discussion

There have been several SEM studies on the normal mucosa of the Eustachian tube in man and animals [3,6,7]. All of these reports have indicated that the mucosa of this organ is more or less a modified ciliated respiratory epithelium with secretory properties. The present investigation has revealed that the distribution and the density of the ciliated epithelial cells in the normal human Eustachian tube varied from place to place. In general, the density of the ciliated cells decreased from the pharyngeal to the tympanic

Fig. 8. Compound cilia are somewhat frequently found in the inflammed tubal mucosa. These may be possibly produced by fusion of several cilia. ×333.

orifice and from the tubal floor to the roof. Therefore, the ciliated epithelial cells on the tubal floor and lower parts of the walls undoubtedly play an important role in active mucociliary mechanism as in other parts of the upper respiratory system. The fine structure of the ciliated epithelial cells in the tubal mucosa was basically similar to that described in the tracheal mucosa [11]. Most of the normal ciliated cells bore approximately 100–200 cilia with an average length of about 8 μm. Their number and length had a tendency to reduce toward the tympanic orifice of the Eustachian tube.

Short cilia-bearing cells were occasionally found. However, no other ciliary abnormalities were noted in the ciliated epithelial cells of the normal tubal mucosa. Goblet cells were interspersed between the ciliated cells and appeared more numerous on the floor and lower one half of the walls than on the vault. The ratio of ciliated cells to goblet cells was approximately 6:1. Even in the normal state of the tubal mucosa, goblet cells showed various processes of mucus secretion. *Sade* et al. [12] have mentioned that mucus is a necessary intermediary in the transport system and that a minimal amount must be available to allow the cilia to perform their transport function.

Fig. 9. Elongated cilia-bearing cells are sometimes found. These giant cilia are two or three times longer than normal ones. × 233.

The inflamed mucosa of the upper respiratory tract is characterized by intercellular edema, infiltration of inflammatory cells and loss of the normal ciliary complement [2]. There have been several reports on the morphological changes of the ciliated epithelial cells of animal and human respiratory mucosa under various experimental and non-experimental conditions. *Duncan and Ramsey* [1] have reported that under the influence of bacterial inflammation, the ciliated epithelial cells of the pig nasal mucosa often became polyhedral in shape. The principal changes were in the cilia, which were reduced in numbers. They have also observed that the markedly enlarged portions of the ciliary shafts occurred at the cell surface. *Von Mecklenburg* et al. [8] have described the presence of the same abnormal cilia in the ciliated cells of the rabbit tracheal mucosa after heat exposure. Various ciliary abnormalities have also been described in association with neoplasms and inflammatory states in animals and man [1, 2].

The present investigation has revealed that abnormal ciliary structures such as short cilia, compound cilia, swollen cilia and elongated cilia were frequently found in the ciliated cells of the inflamed tubal mucosa. A de-

Fig.10. Most giant cilia are thinner than normal ones. Bead-like cilium (arrows) is also found. × 6700.

crease in the number as well as a loss of cilia have also been noted in the ciliated epithelial cells of the diseased tubal mucosa.

Møller and Dalen [9] have observed that some of the ciliated epithelial cells in the presence of middle ear effusion had large numbers of microknobs on the ciliary surface, especially at the tip and upper part of the cilia. *Møller and Dalen* [10] have also mentioned that in the presence of mucoid middle ear effusions more than 10% of all cilia were abnormal. According to them compound cilia and knob-formation of the cilia were the principal changes in the inflammed ciliated cells. In the present investigation on the inflammed mucosa, the most frequent abnormal finding was the lack of coordination of the direction of the cilia. Normal cilia appeared one directional in orientation; however, the orientation of the pathological cilia was random. Pathological cilia were also spaced further apart than those cilia of normal samples. Approximately 10–20% of all cilia showed abnormal figures in the presence of serous middle ear effusions.

Short cilia were sometimes noted in the normal tubal epithelium. However, they were more frequently noted in the inflamed mucosa in the presence of middle ear effusions. Compound cilia and knob- or club-like, swollen cilia have been reported in other respiratory epithelia under various pathological conditions [1, 2, 8]. *Kawabata and Paparella* [5] have found these atypical cilia in the normal human and guinea pig middle ear mucosa. However, such atypical cilia could not be detected in the normal tubal mucosa. Therefore, atypical cilia appear to be more frequent in the diseased or inflamed respiratory type epithelium.

It is the author's opinion that morphological changes of the cilia usually occur prior to those of the cytoplasma. However, until the true nature of these abnormal cilia is elucidated, the possibility remains that they may be a degenerative phenomena. It is reasonable to assume that abnormal cilia impede the regular rhythmic action of the mucociliary apparatus.

Accordingly, an accumulation of fluid easily occurs in the middle ear cavity under such a circumstance. The morphological findings in serous middle ear effusions seem to be unrelated to age. However, further investigation is necessary to clarify the true cause of inducing abnormal cilia in the mucosa of the Eustachian tube and middle ear in various pathological conditions.

References

1 Duncan, J.R.; Ramsey, F.K.: Fine structural changes in the porcine nasal ciliated epithelial cell produced by *Bordelella bronchiseptica* rhinitis. Am. J. Path. *47:* 601–612

2 Friedmann, I.; Bird, E.S.: Ciliary structure, ciliogenesis, microvilli. Laryngoscope, St. Louis *81:* 1852–1868 (1971).

3 Harada, Y.: Scanning electron microscopic study on the distribution of epithelial cells in the Eustachian tube. Acta oto-lar. *83:* 284–290 (1977).

4 Hiraide, F.; Inouye, T.: The fine surface view of the human adult Eustachian tube. J. Laryngol. (in press).

5 Kawabata, I.; Paparella, M.M.: Atypical cilia in normal human and guinea pig middle ear mucosa. Acta oto-lar. *67:* 511–515 (1969).

6 Lim, D.J.: Functional morphology of the lining membrane of the middle ear and Eustachian tube. Ann. Otol. Rhinol. Lar. *83:* suppl. 11, pp. 5–26 (1974).

7   Lim, D.J.: Normal and pathological mucosa of the middle ear and Eustachian tube. Clin. Otolaryngol. *4:* 213–234 (1979).

8   Mecklenburg, C. v.; Mercke, U.; Håkansson, C.H.; Toremalm, N.G.: Morphological changes in ciliary cells due to heat exposure. A scanning electron microscopic study. Cell Tiss. Res. *148:* 45–56 (1974).

9   Møller, P.; Dalen, H.: Middle ear mucosa in cleft palate children. A scanning electron microscopic study. Acta oto-lar. suppl. 360, pp. 198–203 (1979).

10 Møller, P.; Dalen, H.: Ultrastructure of the middle ear mucosa in secretory otitis media. Acta oto-lar. *91:* 95–110 (1981).

11 Rhodin, J.A.G.: Ultrastructure and function of the human tracheal mucosa. Am. Rev. resp. Dis. *93:* 1–15 (1966).

12 Sade, J.; Eliezer, N.; Silberberg, A.; Nevo, A.C.: The role of mucus in transport by cilia. Am. Rev. resp. Dis. *102:* 48–52 (1970).

13 Tanaka, K.: A simple type of apparatus for critical point drying method. J. Electr. Microscop. *21:* 153–154 (1972).

F. Hiraide, MD, Department of Otolaryngology, National Defense Medical College, Saitama (Japan)

Adv. Oto-Rhino-Laryng., vol. 31, pp. 85–117 (Karger, Basel 1983)

The Canaliculae Perforantes of Schuknecht

D. J. Lim, H.-N. Kim

Otological Research Laboratories, Department of Otolaryngology,
Ohio State University College of Medicine, Columbus, Ohio, USA

Introduction

During the course of a histochemical investigation to localize acetyl-cholinesterase, *Schuknecht* fortuitously discovered a system of channels (perilymphatic channels or canaliculae perforantes) that connect the scala tympani with the fluid (cortilymph) bathing the organ of Corti in cats [26–28]. The perilymphatic openings of these channels are pores scattered near the marginal lip of the inferior shelf of the osseous spiral lamina (OSL) [15, 30]. Biochemical analysis of the cortilymph suggests that the fluid bathing the organ of Corti is similar to perilymph in its electrolyte composition, making *Schuknecht's* concept of canaliculae perforantes a plausible explanation of the fluid communication between the cortilymph and perilymph in the scala tympani [22].

Tracer studies demonstrated that the basilar membrane is permeable to macromolecules [6, 10–12, 16, 21, 32]. This raises a question regarding the significance of Schuknecht's canaliculae perforantes as a primary means of fluid communication between the cortilymph and the perilymph. Since clarification of the mechanisms of supplying oxygen and energy to the organ of Corti and removing carbon dioxide from it is vitally important for understanding the physiology and pathology of the hearing organ, the present study was undertaken to further define the communication routes between perilymph and cortilymph by means of light microscopy, electron microscopy (EM), and macromolecular tracer studies.

Materials and Methods

Morphological Study. For the light microscopy study temporal bones from 15 cats, 15 chinchillas, and 15 squirrel monkeys that had been fixed in Heidenhain's susa, decalcified in trichloroacetic acid, embedded in celloidin, and cut in 20-μm thick sections were used. In addition, five chinchilla cochleas were fixed in 2% glutaraldehyde with cacodylate buffer, decalcified in 4.5% EDTA for 2 weeks, embedded in JB-4 plastic embedding medium (Polysciences), and serially sectioned with glass knives at a thickness of 5 μm. These celloidin and JB-4 sections were stained with hematoxylin and eosin, examined under a microscope fitted with a drawing tube, and tracings made of the regions of interest. The distance between the margin of the osseous spiral lamina and the habenula perforata and the location and sizes of the bony pores in the OSL were measured from each drawing using a digitizer (Hewlett-Packard).

For transmission EM five normal chinchilla cochleas were used after being fixed in 2% glutaraldehyde with cacodylate buffer (some specimens were decalcified in 4.5% EDTA – 2% glutaraldehyde), postfixed in 1.33% osmic acid buffered with *s*-collidine, and embedded in epoxy resin.

For scanning electron microscopy (SEM) five chinchilla cochleas were fixed in 2% phosphate-buffered osmic acid and dissected in ethanol. They were embedded in epoxy resin and stored overnight in a freezer to prevent the polymerization of the epoxy. The frozen cochleas were fractured with cold razor blades on a cold stage, as described by *Tanaka* [29]. After thawing, they were washed overnight in propylene oxide (several changes) to remove the epoxy resin, and the washed specimens were critical point dried in CO_2, gold coated using a sputter coater (Polaron SEM Coating Unit E5100), and examined under the scanning electron microscope (Cambridge Mark IIa). Two additional chinchilla temporal bones were processed to remove the cells covering the undersurface of the inferior OSL. These glutaraldehyde and osmic acid fixed specimens were emersed in 40% KOH overnight, rinsed in distilled water, dehydrated in graded ethanol, critical point dried in CO_2 and gold coated before being examined under SEM.

Tracer Study. The tracers used were horseradish peroxidase (HRP; Sigma type II, MW 33,000; Sigma) and horse spleen ferritin (Nutritional Biochemical). The ferritin was dialyzed to remove any residual cadmium, as described by *Hinojosa* [9], and checked with an X-ray analyzer for purity. The stock solution was sterilized by passing it through a 0.45-μm mesh Millipore filter and refrigerated in sealed ampules.

Under anesthesia (Metofane), both ears of 12 chinchillas were stapedectomized. Through the round window membrane 0.1 ml of ferritin or HRP suspension (100 mg HRP/1 ml Ringer's solution) was slowly infused through a fine capillary pipet, connected to a 2-ml syringe with polyethylene tube and mounted on a Harvard infusion pump (six ears for HRP, six as controls for HRP, twelve for ferritin). The animals were sacrificed at approximately 5, 15, or 30 min or 1 h following introduction of the tracers. The HRP-injected specimens were prepared in the manner described by *Karnovsky* [13]. Both sets of specimens (ferritin and HRP) were decalcified in either 4.5% EDTA or 4.5% EDTA – 2% glutaraldehyde for 2–3 weeks, embedded in epoxy resin, and stained with uranyl acetate. Some specimens were embedded without decalcification.

Results

Morphology. The inferior shelf of the OSL was wider in the basal turn and narrower in the apical turn (table I). Numerous bony pores were found in the lower shelf of the OSL near its outer margin in all animals examined (table II), and their average diameters under light microscopy were 7.1 μm in chinchillas, 13 μm in cats, and 10 μm in squirrel monkeys. The smallest one observed under EM in a chinchilla was 0.25 μm in diameter and the largest was 10 μm. Although they sometimes were loosely covered by spindle-shaped perilymphatic cells, they often remained open (fig. 1, 2). These pores were found along the entire length of the cochlea, and their density increased from the base to the apex (table III). The pores were clustered near the outer half (organ of Corti side) of the OSL, forming a band of perforated inferior bony shelf called the zona perforata (fig. 1, table III). Blood vessels were found near the pores in 60% of the specimens; some, particularly those in the basal turn, were found on the scala tympani side of the bony shelf (fig. 3). Some capillaries had penetrated the pores, but these were rather rare.

The margin of the lower shelf of the OSL is closer to the habenula perforata in the basal turn than it is in the apex. This results in a large area of bony dehiscence (dehiscent zone) and exposes a large area of myelinated nerve fibers and capillaries to the perilymph (fig. 4, table IV), although this area is well covered by basilar membrane mesothelial cells (fig. 4b).

Very wide perineural channels (perineural spaces and gaps) connecting the bony pores with the habenula perforata were observed in EM (fig. 4b, 5). The habenula perforata consists of regularly arranged holes in the basilar

Table I. Radial length of osseous spiral lamina (μm)[1]

Turn of cochlea	Chinchillas (n = 15)		Cats (n = 15)		Squirrel monkeys (n = 15)	
	n²	mean ± 1 SD (range)	n²	mean ± 1 SD (range)	n²	mean ± 1 SD (range)
Basal	15	422.2 ± 24.90 (388–458)	15	507.1 ± 33.22 (465–572)	15	641.1 ± 43.90 (514–700)
Middle	15	310.4 ± 21.86 (267–365)	15	350.4 ± 26.53 (319–419)	15	485.7 ± 59.41 (379–608)
Apical	15	166.5 ± 23.79 (133–205)	15	229.5 ± 30.86 (167–300)	15	218.3 ± 24.58 (170–260)

[1] Radial length is defined as the distance between the margin of the OSL and the outer margin of Rosenthal's canal containing ganglion cells.
[2] Number of midmodiolar sections.

Table II. Size of bony pores (μm) as measured by light microscopy

Turn of cochlea	Chinchillas (n¹ = 5)		Cats (n¹ = 4)		Squirrel monkeys (n¹ = 5)	
	n²	mean ± 1 SD (range)	n²	mean ± 1 SD (range)	n²	mean ± 1 SD (range)
Basal	47	6.7 ± 2.90 (1.2–13.7)	37	12.4 ± 5.41 (2.3–22.1)	152	10.2 ± 4.91 (1.6–26.7)
Middle	42	7.0 ± 2.51 (2.3–12.8)	32	14.6 ± 7.11 (1.6–29.1)	124	10.2 ± 5.77 (1.2–32.6)
Apical	36	7.7 ± 4.21 (1.4–23.3)	17	11.2 ± 5.24 (1.2–25.6)	43	8.1 ± 3.48 (2.3–15.1)
Average	125	7.1 ± 3.23 (1.2–23.3)	86	13.0 ± 6.24 (1.2–29.1)	319	9.9 ± 5.15 (1.2–32.6)

[1] Number of cochleas.
[2] Number of bony pores measured.

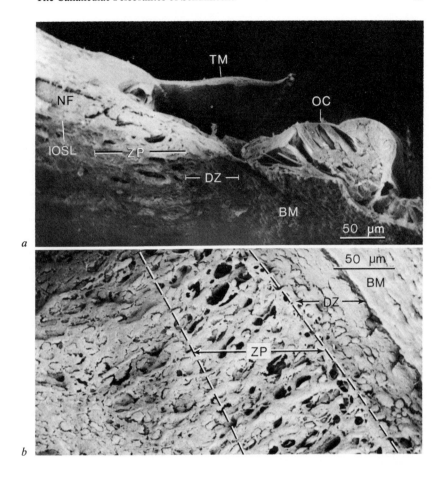

Fig. 1. a SEM photomicrograph of Epon-embedded and freeze-fractured chinchilla cochlea (apical turn) shows the undersurface of the inferior shelf of the osseous spiral lamina (IOSL) with zona perforata (ZP), dehiscent zone (DZ), and basilar membrane (BM). The largest pore measures about 30 μm. NF = Nerve fibers; TM = tectorial membrane; OC = organ of Corti. *b* SEM photomicrograph shows the undersurface of the OSL with zona perforata (ZP) in a chinchilla basal turn. The dehiscent zone (DZ) is covered by mesothelial basilar membrane cells. BM = Basilar membrane.

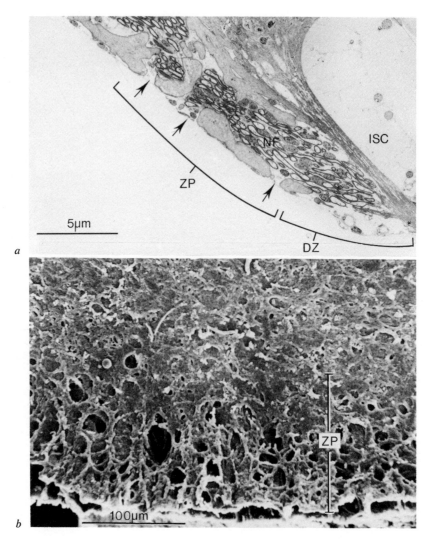

Fig. 2. a TEM photomicrograph shows inferior OSL of the second turn of a chinchilla cochlea with numerous bony pores (arrows) in the zona perforata (ZP). Nerve fibers (NF) in the spiral lamina are also exposed to the perilymph at the dehiscent zone (DZ). ISC = Inner sulcus cells. *b* SEM view of the undersurface of the OSL at the second turn of a chinchilla after KOH treatment to remove mesothelial cell covering. Large pores are clustered in the zona perforata (ZP), and small pores are scattered over the rest of the bony shelf.

Table III. Density of bony pores per 0.1 mm² of inferior shelf of osseous spiral lamina

Turn of cochlea	Chinchillas (n¹ = 5)		Cats (n¹ = 4)		Squirrel monkeys (n¹ = 5)	
	outer half	inner half	outer half	inner half	outer half	inner half
Basal	59.9 ± 12.88	14.1 ± 7.98	37.7 ± 2.54	10.4 ± 2.92	102.7 ± 3.84	46.1 ± 4.08
Middle	81.1 ± 18.54	26.1 ± 15.45	46.6 ± 5.29	10.8 ± 3.64	124.9 ± 22.0	46.6 ± 21.82
Apical	122.6 ± 21.42	48.4 ± 18.56	71.1 ± 27.9	21.9 ± 0.64	161.2 ± 65.56	91.9 ± 51.08

¹ Number of cochleas.

Table IV. Width of dehiscence of spiral lamina (μm)¹

Turn of cochlea	Chinchillas (n² = 5)		Cats (n² = 4)		Squirrel monkeys (n² = 5)	
	n³	mean ± 1 SD (range)	n³	mean ± 1 SD (range)	n³	mean ± 1 SD (range)
Basal	14	34.6 ± 7.30 (22–46)	15	40.2 ± 23.62 (8–88)	16	44.4 ± 21.21 (16–79)
Middle	13	45.5 ± 11.93 (35–79)	16	84.5 ± 20.86 (42–116)	15	119.9 ± 24.71 (92–178)
Apical	13	119.5 ± 21.88 (91–166)	8	162.5 ± 10.47 (152–186)	8	166.4 ± 28.02 (144–230)

¹ The distance between the outer bony margin of the inferior OSL and the habenula perforata.
² Number of cochleas.
³ Number of midmodiolar sections examined.

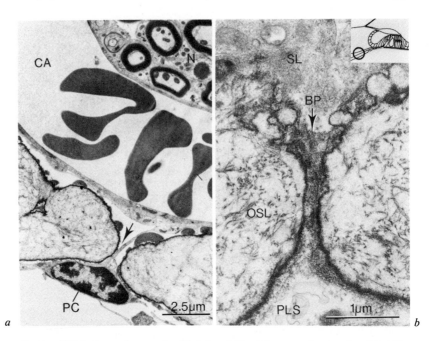

Fig.3. a TEM photomicrograph of inferior OSL of the second turn of a chinchilla cochlea showing a small bony pore (arrow) that is covered by a perilymphatic cell (PC). A capillary (CA) is near the bony pore. N = Nerve fibers. b Small bony pore (BP) in the inferior OSL of the second turn of a chinchilla cochlea is filled with ferritin particles 10 min after the tracers were introduced into the perilymph through the round window membrane. SL = Spiral laminar space; PLS = perilymphatic space.

membrane through which the fibers penetrate (fig. 6). Just before penetration the myelinated nerve fibers become unmyelinated, and there are perineural fluid gaps and spaces around the bundles of the penetrating bare nerve fibers (fig. 4b, 5, 7). There are also small fluid spaces surrounding the penetrating nerve fiber bundles, which should allow free communication of perilymph with the perineural spaces of the organ of Corti (fig. 8–10). The perineural gaps measured about 300–460 Å in width, and dilated perineural or interneural spaces measured from 0.1 to 0.5 μm in width (fig. 10).

The cell junctions of the basal portion of the supporting cells lining the basilar membrane are macular-type gap junctions (macula adherentes), the widths of which measured about 20 Å (fig. 11), in contrast with the supporting cell apices, which are sealed by the zonula occludens and zonula adherens

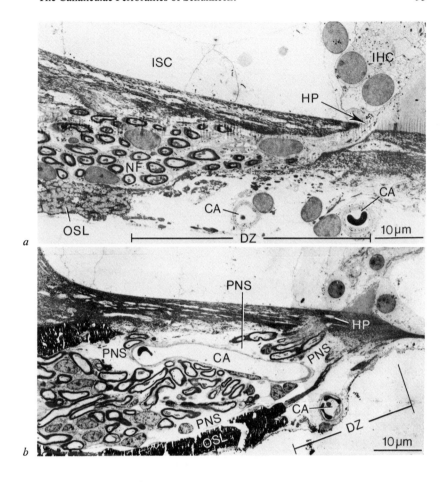

Fig.4. a TEM photomicrograph shows a zone of large (about 50 μm wide) dehiscence (DZ) of inferior OSL of a chinchilla apical turn. Spiral vessels (CA) are exposed to perilymph. HP = Habenula perforata; IHC = inner hair cells; ISC = inner sulcus cells; NF = nerve fibers. *b* Zone of small (25 μm wide) dehiscence (DZ) of inferior shelf of OSL of a chinchilla apical turn is shown. The dehiscent area is lined with perilymphatic cells, the outline of which is clearly marked by HRP that had been injected into the scala tympani 14 min earlier. Observe large perineural spaces (PNS) and pericapillary space in the spiral lamina. HP = Habenula perforata; CA = capillary.

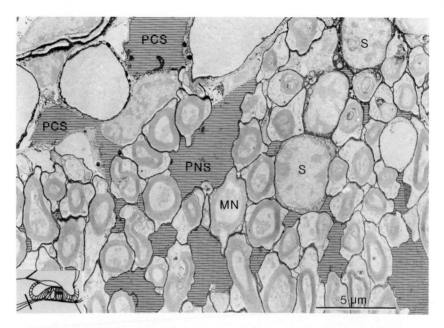

Fig. 5. TEM photomicrograph shows cross-sectional view of unstained bundles of myelinated nerve fibers (MN) very near to the habenula perforata in basal turn. Dark peripheral lines encircling the nerves are caused by deposition of HRP 5 min after tracer instillation via the scala tympani. Numerous large perineural (PNS) and pericapillary (PCS) spaces are marked with parallel lines. These spaces represent about 14–18% of the nerve areas measured in the photomicrograph. S = Schwann's cells.

(fig. 12). Basement laminae were found along the basal surfaces of the supporting cells in contact with the basilar membrane, even when a large gap was created by the separation of two adjoining cells (fig. 13). The absence of the basement lamina in the basilar membrane observed in the cat by *Cabezudo* [4] was not found in the chinchilla.

Fig.6. a Low-power view of the habenula perforata (inside the rectangle) that is formed by a tubular hole in the basilar membrane fibers. In this specimen the habenula perforata appears open because of the missing nerve fibers (unrelated to the experiment). IH = Inner hair cell; BC = border cell; IP = inner pillar cell; ISC = inner sulcus cell; IPh = inner phalangeal cell. The rectangle is magnified in *b*. *b* Close-up view from *a* showing the details of a habenula perforata (HP) that consists of a linearly arranged hole in the basilar membrane through which unmyelinated nerve fibers penetrate. IP = Inner pillar cell; IPh = inner phalangeal cell; ISC = inner sulcus cell; BMF = basilar membrane fibers.

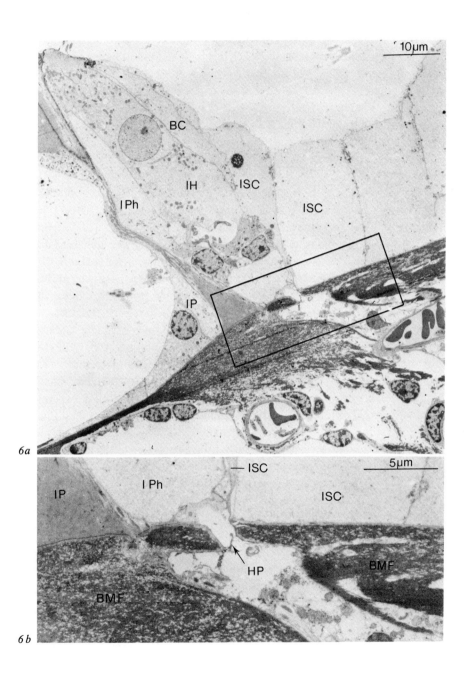

6a

6b

Tracer Study. In the spiral lamina, both HRP and ferritin readily passed
through the bony pores of the OSL, even when the pores were covered by the
perilymphatic cells (fig. 3b), and through the habenula perforata along the
bare nerve fibers (fig. 7–9). We consistently found a larger amount of tracer
in the perineural spaces under the inner hair cells than in other areas of the
organ of Corti, regardless of the tracer type (fig. 14a, b, 15, 16). This result is
interpreted to indicate that they had freely entered Corti's tunnel, mostly via
the perineural spaces and gaps (fig. 7–10). They were also observed along the
inner and external spiral bundles (fig. 17, 18) and the tunnel fibers and cell surfaces (fig. 19), even when the tracers did not penetrate the basilar membrane at
all, as clearly demonstrated in a few specimens and shown in figure 7, 14, 18, 20
and 21.

In some specimens the tracers penetrated the intercellular spaces between
the bases of the pillar cells that form the basal border of Corti's tunnel (fig. 19);
in others they did not. Neither the HRP nor the ferritin penetrated the gap
junctions; however, the tracers can bypass them since the gap junctions do
not completely seal the entire circumferences of the intercellular spaces in the
way the zonula occludens does. A similar intercellular transporting mechanism
was observed in the intercellular spaces of other supporting cells that are in
contact with the basilar membrane: Deiters', Hensen's, Claudius', Boettcher's,
and inner sulcus cells (fig. 22). It was not possible to accurately quantitate the
amount of HRP that had penetrated the basilar membrane because of the
high intensity of enzyme reaction required to visualize the macromolecules
and the aggregation of the tracer. However, when ferritin was used individual
particles could be seen in relatively small amounts in the intercellular spaces
of the supporting cells (fig. 22). This is in contrast with the large amount that
penetrated via the perineural spaces, as mentioned above in the same specimen. Some of the tracer was apparently transported transcellularly by the
vesicular transporting system in the inner sulcus cells and Claudius' cells, but
the number of such vesicles was very small.

Although the tracers that were found in the intercellular spaces of the
supporting cells are assumed to have been transported from the perilymph
side of the basilar membrane, the exact direction of the passage of the particles
could not be determined, since only a small amount (if any) of HRP was found
in the basilar membrane in certain cases (fig. 15, 19). In fact, large amounts
of HRP accumulated in the intercellular spaces of the inner sulcus in rare
cases when there was none in the basilar membrane (fig. 23a). Also, in one
specimen we observed saturation of a discrete area of the basilar membrane
and the intercellular spaces with HRP, giving the impression that the HRP

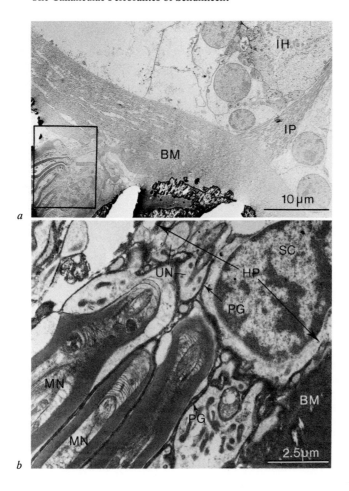

Fig. 7. a Low-power TEM photomicrograph shows a portion of the organ of Corti including inner hair cells (IH), inner pillar cells (IP), basilar membrane (BM), and myelinated fibers near the habenula perforata (in the rectangle). (Specimen No. 0934L.) *b* Close-up view of the rectangle from *a* (but from an adjacent section) shows an entrance to the habenula perforata (HP), where dense demyelinization occurs. Observe the HRP (darkly stained areas) in perineural gaps (PG). MN = Myelinated nerve; UN = unmyelinated fibers; SC = Schwann cell. 14 min after HRP injection in the basal turn.

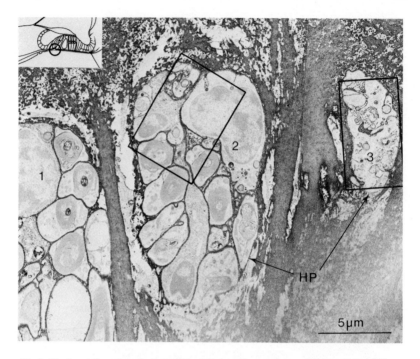

Fig.8. Horizontal sectional view of three habenulae perforate (HP) at three different levels because of the slanted sectioning angle. Nerve fibers in 1 and 2 are still myelinated and in 3 are unmyelinated. The animal was sacrificed 5 min after HRP injection into the scala tympani. The section was unstained to show HRP localization, which is shown in 2 and 3. The 2 and 3 rectangles are magnified in figure 9.

may have been transported from the organ of Corti side to the basilar membrane side (fig. 23b).

In general, the tracer particles were confined within this cortilymph space and did not pass through the zonula occludens of the junctional complexes of the epithelial cells (fig. 22). However, a very small amount of ferritin appeared to have been transported across the epithelial cells to the endolymph via reverse pinocytosis (fig. 24).

Many vessels found near the dehiscent area of the OSL appeared to be venous, as they took up HRP particles, in contrast to the arterial vessels in the OSL, which failed to take up the tracers. The penetration of the tracers through the basilar membrane is time and thickness dependent. The basilar membrane is very thick in the basal turn (about 5 μm) and thin (about 1 μm) in the apical turn in chinchillas. Therefore, the basilar membrane in the apical turn is more

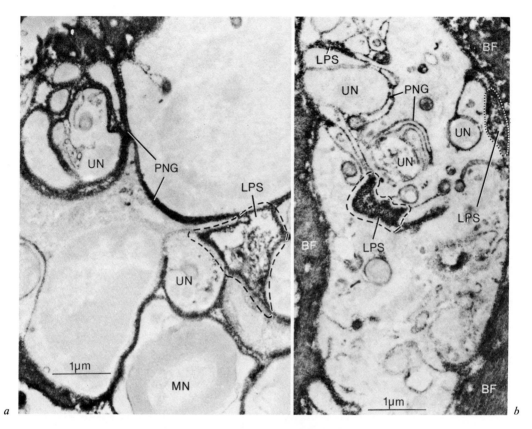

Fig. 9. a Magnified view of the No. 2 habenula in figure 8 shows packed HRP reaction products in perineural gaps (PNG) and a large perineural space (LPS) among unmyelinated (UN) and myelinated nerve (MN) fibers. The large cells with nuclei are interpreted as being Schwann cells. 5 min post-HRP injection, basal turn. *b* A magnified view of the No. 3 habenula in figure 8 also shows numerous perineural gaps (PNG) and a few large perineural spaces (LPS) among many unmyelinated fibers (UN). The large cells are interpreted as being supporting cells. 5 min post-HRP injection, basal turn. BF = Basilar membrane fibers.

readily permeable than that in the basal turn. In the 5-min HRP specimens, there was evidence of slow diffusion in the basal turn (fig. 20); however, in the 30-min specimens the basilar membrane ground substance became saturated with HRP. It is important to note that the HRP often appeared in a clump in the basilar membrane (fig. 20), suggesting uneven distribution of tracer along the length of the cochlea. This phenomenon was not obvious when ferritin was used.

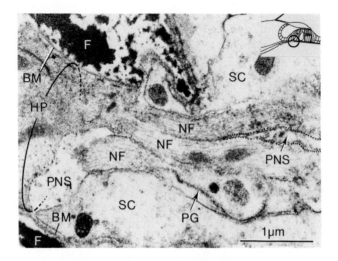

Fig. 10. TEM photomicrograph of the habenula perforata (HP) in the apical turn of a chinchilla cochlea with passing unmyelinated nerve fibers (NF) and large and small perineural spaces (PNS) containing HRP particles 14 min. post-injection via scala tympani. HRP reaction products are seen in perineural gaps (PG) and spaces. BM = Basal membrane; F = basilar membrane fibers; SC = supporting cells.

We have examined the possibility of cellular toxicity caused by the tracers. No evidence of neural or cell damage was seen when ferritin was used. However, on rare occasion, swelling of unmyelinated nerves was observed in specimens in which HRP was used. In 1 case in which the basilar membrane was completely saturated with HRP, the tracer also penetrated the supporting cells of the organ of Corti, causing severe cell damage.

Discussion

The origin of perilymph and how oxygen and nutrients are supplied to the organ of Corti have been focal points of research interest. Based on a series of experiments, *Misrahy* et al. [17, 18] suggested that the oxygen supply to the sensory cells derives mainly from the endolymph, although small amounts might be supplied to the sensory cells from the scala tympani. On the other hand, *Lawrence and Nuttall* [14] demonstrated that oxygen supplying the sensory cells is most likely derived mainly from cortilymph which, in that case, would receive its oxygen supply from the spiral vessels. *Schnieder* [25]

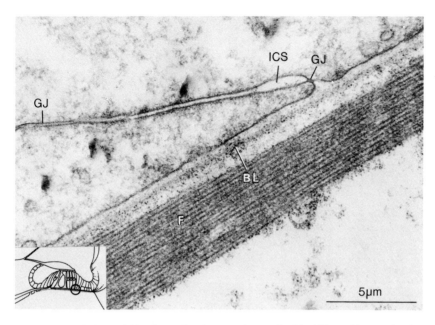

Fig. 11. The base of Claudius' cell at the second turn of a chinchilla cochlea 10 min after ferritin injection via the round window. A large number of ferritin particles are trapped along the basal lamina (BL), and only a few passed through the membrane into the intercellular spaces (ICS). F = Basilar membrane fibers; GJ = gap junction.

calculated the oxygen supply to the organ of Corti via perilymph to be 0.1 nM min, whereas the oxygen supply via endolymph was only 1.6% of the perilymphatic supply. *Tasaki and Fernández* [31] were able to abolish cochlear microphonics and action potentials by perfusing the scala tympani with artificial endolymph. When the scala vestibuli is perfused with the same solution, there is a time lapse before the microphonics are depressed. This phenomenon is attributed to the diffusion of perilymph (which carries oxygen and nutrients) from the scala vestibuli toward the scala tympani.

If perilymph communicates with cortilymph, as implied in the above observations, one may wonder whether there is a morphologic basis for the fluid passage between these two fluid systems. The discovery of 'perilymphatic channels' by *Schuknecht* [26] gave credence to the idea that there may be distinct anatomic channels connecting the cortilymph with the perilymph in the scala tympani. However, previous ultrastructural studies have failed to substantiate any remarkable perilymphatic channels [7,8]. Bony pores that

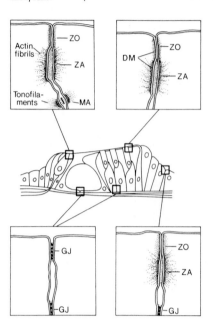

Fig.12. Schematic diagram of various specialized cell junctions observed in the chinchilla organ of Corti. The apical parts of cells are sealed by a junctional complex formed of the zonula occludens (ZO), zonula adherens (ZA), and macula adherens (MA), known as a desmosome. The latter sometimes is missing. Dense materials (DM) are also found adjacent to the zonulae adherens and occludens. The basal portions of all supporting cells are attached to each other only by gap junctions (GJ).

are similar to those *Schuknecht* described in cats are present in guinea pigs, as reported earlier by us and others [15, 24, 30], and the present findings indicate that they are also present in chinchillas and squirrel monkeys.

Numerous tracer studies have established the concept that perilymph freely communicates with cortilymph, but there are some differences of opinion about the exact route. *Von Ilberg* [10], using Thorotrast, and *Duvall* and *Sutherland* [6], using HRP, demonstrated that the tracer particles readily enter the cortilymph spaces through the perineural spaces of the habenula perforata and also by diffusion through the basilar membrane and via the intercellular spaces. *Asher and Sando* [2], using HRP, observed that most HRP entered the organ of Corti via the intercellular spaces and, to a lesser degree, via the habenula perforata. On the other hand, *De Lorenzo* et al. [5], using micro and macro HRP particles, confirmed that the habenula perforata is the route of

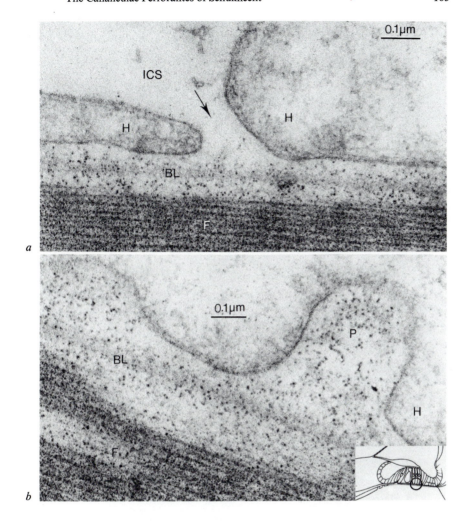

Fig. 13. a High-power electron micrograph shows a large gap (arrow) between two Hensen's cell (H) bases. Observe ferritin particles that are found in the basal lamina (BL) and in the basilar membrane, including the ground substance (not shown) and fibers (F). Very little ferritin is found in the intercellular space (ICS). 10 min after ferritin injection. *b* High-power electron micrograph of the base of a Hensen's cell (H) shows an inpouching of the cell, creating a pocket (P). A heavy accumulation of ferritin is seen in this pocket, suggesting the ferritin can permeate through the basilar membrane. 10 min after ferritin injection. Ferritin also penetrated the basilar membrane fibers (F). BL = Basal lamina.

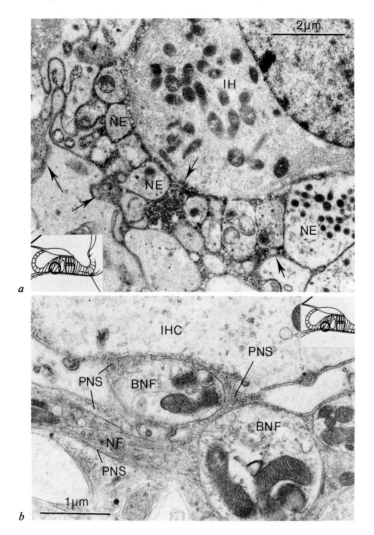

Fig. 14. a Heavy accumulation of HRP in the peri- and interneural spaces (arrows) under an inner hair cell (IH). HRP is seen between nerve fibers as well as between nerve endings (NE) and hair cells. 10 min post-HRP injection. *b* Densely packed ferritin particles are found in the perineural spaces (PNS) containing unmyelinated nerve fibers (NF) and bulbous bare nerve fibers (BNF) under the inner hair cells (IHC) in the apical turn of a chinchilla cochlea 20 min after introduction of tracers via the round window.

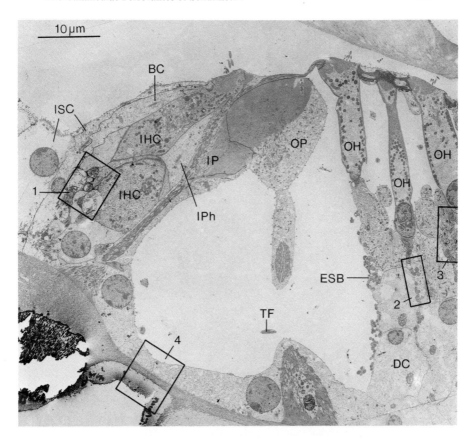

Fig. 15. Low-power electron micrograph of the organ of Corti of a chinchilla (0934L) that received HRP via the scala tympani 14 min before sacrifice. Heavy accumulation of HRP is seen under the inner hair cells (IHC), shown in rectangle 1. Some HRP has also accumulated along the external spiral bundles (ESB), shown in rectangle 2, and along the nerve ending under the outer hair cells (OH), shown in rectangle 3, in the absence of HRP penetration through the basilar membrane (rectangle 4). ISC = Inner sulcus cell; IPh = inner phalangeal cell; BC = border cell; IP = inner pillar cell; OP = outer pillar cell; DC = Deiters' cell; TF = tunnel fiber.

perilymph-cortilymph communication, but they failed to show any passage through the basilar membrane. *Masuda* et al. [16] demonstrated that when radioisotope-labeled inulin was injected into the scala tympani of guinea pigs a heavy accumulation of the labeled substance appeared in the habenula perforata, in the lacunae of Corti (cortilymph spaces), and, to a far lesser degree, in the intercellular spaces of Hensen's and Claudius' cells [ref. 16,

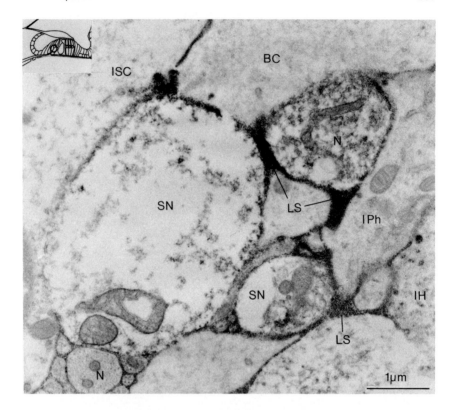

Fig. 16. Close-up view of rectangle 1 from figure 15 showing heavy accumulation of HRP in the perineural spaces near the inner hair cells (IH). N = Nerve fibers or nerve endings; SN = swollen nerve fibers; ISC = inner sulcus cell; IPh = inner phalangeal cell; BC = border cell; LS = large interneural spaces.

fig. 1, 2]. Their findings clearly indicate that the isotope-labeled inulin mainly passed through the habenula perforata route and may have passed through the basal portion of the pillar cells. The present data from both HRP and ferritin are in agreement with the findings reported by *Masuda* et al. [16].

Interpreting the results of tracer studies is not simple because one must consider toxicity, particle charges and size, and other biological behavior. We were particularly concerned about the toxicity. *Ross* et al. [23] reported acute signs of sensory cell damage (swelling and vacuolization) following cochlear perfusion with 1% HRP in artificial perilymph. We observed some evidence of cell and nerve injury in an occasional HRP specimen, but most of the cases did not show any pathologic changes to indicate overt toxicity.

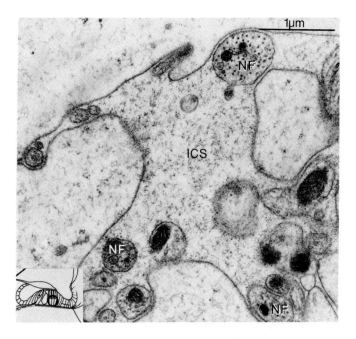

Fig.17. The specimen shown in figure 14b in the apical coil of a chinchilla cochlea 20 min after ferritin injection. Loosely packed ferritin particles are found in the perineural space (ICS) of the external spiral fibers (NF) under the outer hair cells, in contrast with the dense ferritin packing in the perineural spaces under the inner hair cell shown in figure 14b.

Even assuming that the swelling of the nerves was caused by the HRP, the swelling would impede the passage of tracer, not facilitate it. Therefore, we consider our present results valid. Furthermore, we did not see any evidence of cell or nerve injury from ferritin, suggesting that the pure ferritin we used is biologically inert. These data are in agreement with those obtained using HRP.

The size of the tracers or molecules is also important for understanding the diffusion mechanisms across and between the epithelial cells. Many tracers of varying sizes have been used in defining fluid spaces and mechanisms of fluid and macromolecular transport in the cochlea. Particle sizes, or molecular weight, of tracers that have been used in inner ear studies are as follows: (a) inulin – MW 5,000–5,500 [16]; (b) micro-HRP – MW 20,000 [5]; (c) HRP – MW 33,000–40,000 [6]; (d) thorium dioxide – 40–100 Å [10]; (e) ferritin – 50–100 Å [9]; (f) colloidal carbon – 200–400 Å [24]. *Sando* et al. [24] failed to observe any colloidal carbon particles beyond the habenula perforata when

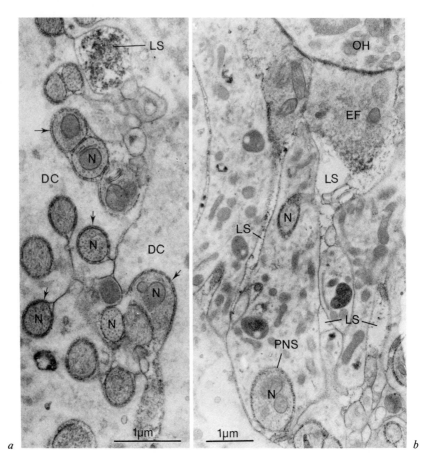

Fig. 18. a Close-up view of external spiral bundles from the specimen shown in figure 15 (some sections away from that shown in rectangle 2). It shows heavy accumulation of HRP in the perineural gap (or space) indicated by arrows. A large intercellular space (LS) is created by Deiters' cells (DC). N = Unmyelinated nerve fibers. *b* Close-up view of rectangle 3 from figure 15, near the base of the third outer hair cell, shows large intercellular (LS) or perineural (PNS) spaces between the nerve fibers (N) and surrounding supporting cells. There was no evidence of HRP penetration in the basal portion of Deiters' cells or the basilar membrane of this specimen like that shown in figure 21. EF = Efferent fiber; OH = outer hair cell.

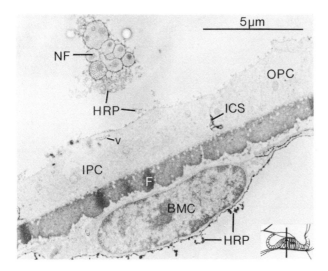

Fig.19. HRP particles are found on the cortilymph surfaces of the inner (IPC) and outer (OPC) pillar cells, in the cytoplasmic vesicles (v) and intercellular spaces (ICS), as well as on the perilymphatic surface of the basilar membrane cell (BMC). F = Basilar membrane fibers; NF = nerve fibers. 4 min after HRP injection via round window, apical turn.

the tracer was introduced into the scala tympani via the cerebrospinal fluid (CSF) space. It is interesting to note that the tracer particles less than 100 Å, which are smaller than carbon particles, all apparently passed through the perineural space of the habenula perforata in large amounts and also, to a lesser extent, through the basilar membrane. This finding can be interpreted as meaning that some of the large perineural spaces observed along the bare nerve fibers in the habenula perforata may not all be interconnected, since carbon particles would have easily passed through if these microchannels of the large perineural spaces are interconnected. It is also possible that the carbon particles we used were larger than 450 Å, or that the perineural spaces seen in the habenula perforata may be exaggerated by the tissue processing. Further study is needed to clarify this question.

The morphology of cell junctions in the organ of Corti is identical to that of other epithelial cells [19, 20], and the results of tracer studies, including ours, are in agreement with the established concept [3]. The macromolecules appear to pass through the intercellular space but fail to pass through the tight junction (zonula occludens) in the organ of Corti, although it has been suggested that micromolecular HRP can pass through tight junctions [5].

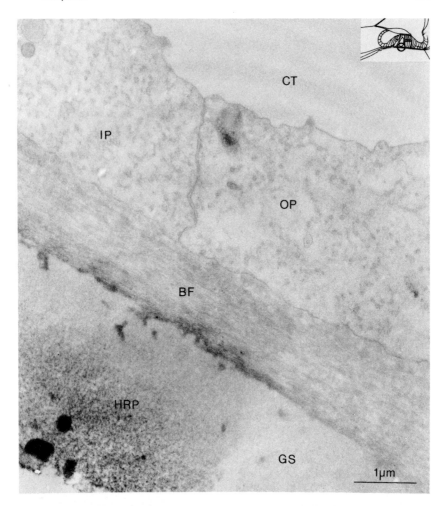

Fig. 20. Close-up view of the basilar membrane and bases of inner (IP) and outer (OP) pillar cells from a chinchilla (0934L) that shows heavy accumulation of HRP all along the perineural spaces, as shown in figures 15, 16, 18, 20, 21. The HRP diffusion is graded and spotty in the ground substance (GS) of the basilar membrane. No HRP was found in the intercellular space of the pillar cells or inside the tunnel of Corti (CT), although it partly diffused through the basilar membrane fibers (BF).

The basal portions of the supporting cells of the organ of Corti are held together by only a few gap junctions of the macular type. The presence of only this junction type indicates that this area may be an important route for macromolecular diffusion between cortilymph and perilymph. Because the

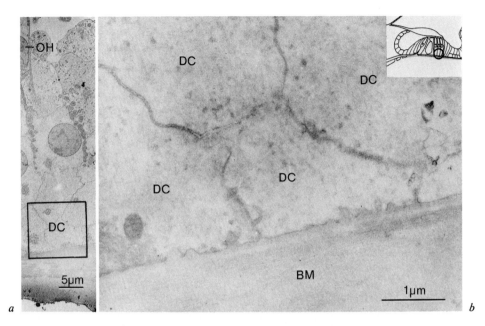

Fig. 21. a Low-power view of external spiral bundles under the outer hair cells (OH) and Deiters' cells (DC) from the area shown in figure 15 but a few sections away. The boxed area is magnified in *b*. *b* In this magnification from *a* there is a trace of HRP in the intercellular spaces of Deiters' cells (DC) but none in the basilar membrane (BM).

basal portions of the inner and outer pillar cells are the thinnest and closest to the spiral vessels compared with the other basal portion of the supporting cells, this tunnel area should be an important area for diffusion across the basilar membrane.

Based on the studies of others and the present study, the basilar membrane is undoubtedly permeable to macromolecules and ions. *Angelborg and Engström* [1] observed that when a small amount (0.05–0.1 ml) of thorium dioxide was injected through the CSF it may be found in the basilar membrane but not in the organ of Corti. They found a diffusion gradient of tracer in the amorphous layer of the basilar membrane, and our data are in agreement with theirs concerning the presence of a diffusion gradient. The results of tracer passage are often difficult to quantitate, particularly when large quantities of tracers are introduced; however, when a small amount of HRP or ferritin was used there appeared to be an indication that passage through the perineural spaces of the habenula perforata is far greater than through the intercellular

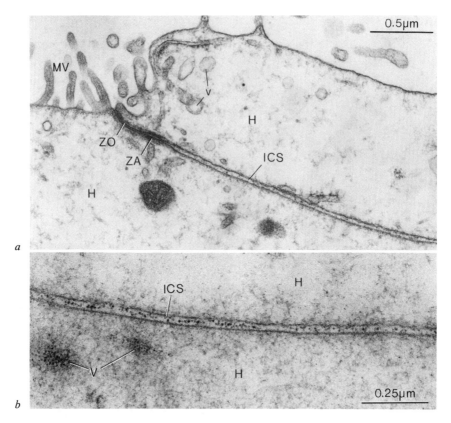

Fig. 22. a The apical portion of Hensen's cells (H) showing a small amount of ferritin in the intercellular space (ICS) 10 min after the tracer was introduced via the scala tympani. No ferritin is seen in the perilymph side of the cell surface. MV = Microvilli; ZO = zonula occludens; ZA = zonula adherens; v = vesicle. *b* High-power view of a similar area of the same specimen as in *a*. Ferritin is also seen inside the cytoplasmic vesicles (V) and in the intercellular space (ICS). H = Hensen's cells.

spaces across the basilar membrane. This observation is similar to results obtained by *Masuda* et al. [16] using inulin.

The exact mechanism by which ions and water are transported between the scala tympani and the cortilymph in a disturbed state is not fully known, although it is assumed that an exchange of solutes must occur between these two compartments. Diffusion by gradient differences is the most likely mechanism involved. In this light we have considered the functional significance of the zona perforata and dehiscent zone that are found in the lower bony

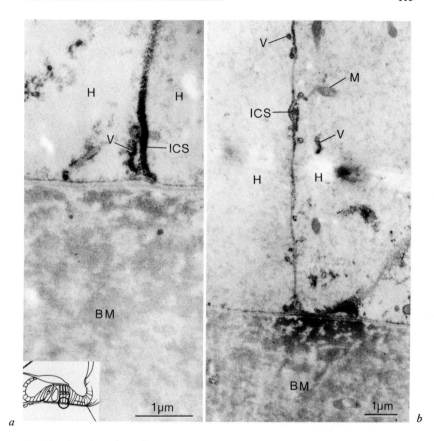

Fig. 23. a Basal portion of Hensen's cells (H) showing heavy HRP accumulation in the intercellular space (ICS) and in the vesicle (V), in contrast with the absence of HRP in the basilar membrane (BM). 14 min after HRP introduction via the scala tympani. *b* Same specimen as *a* but a different section, in which the tracer appeared to diffuse out from the intercellular space (ICS) of Hensen's cells (H) to the basilar membrane (BM). V = Vesicle; M = mitochondria.

shelves of the spiral lamina. We found over 60% of the bony pores of the osseous spiral lamina associated with capillaries in all the species examined, in agreement with *Sando* et al. [24]. Oxygen and serum protein that are transudated may therefore have quick access to the perilymph underneath the organ of Corti. This idea is attractive because most of the arterial vessels that are near the organ of Corti are housed inside the OSL, and its bony pores would allow perilymph to be in direct contact with the arterial vessels. This arrangement would be particularly important in supplying oxygen to the

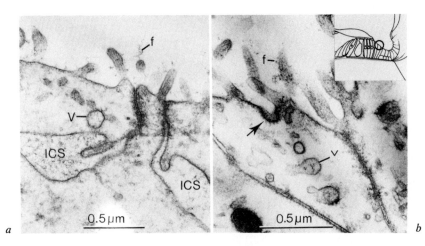

Fig. 24. a Apical portion of Claudius' cells in the apical turn of a chinchilla cochlea shows well-developed junctional complex and large intercellular spaces (ICS) containing some ferritin particles. A cytoplasmic vesicle (V) also contains some ferritin. A few ferritin particles (f) are seen on the surfaces of microvilli. 20 min post-ferritin injection via round window membrane. *b* Same as *a*. Some ferritin particles were expelled into the endolymphatic space by the vesicular transporting system (v) and via exocytosis (arrow). A few ferritin (f) particles are attached to the surfaces of the microvilli.

organ of Corti because the spiral vessels in the basal turn in certain species are known to degenerate after the maturation of the cochlea.

Spiral vessels that contain large amounts of HRP may be venous, and other vessels that do not may be arterial. This could make a very effective system to deliver oxygen and nutrients to the perilymph closest to the organ of Corti and at the same time remove wastes from it. In this regard, it is interesting that the large bony dehiscence of the inferior OSL is greater in the apical turn than in the basal turn. Together with the numerous bony pores in the zona perforata, this dehiscence of the OSL may be an added anatomic arrangement to ensure rapid diffusion of O_2, CO_2, and nutrients between the perilymph and cortilymph bathing the sensory cells.

Another possibility is that the arterial blood vessels may have a pulsation that acts as a pump to facilitate perivascular and perineural fluid forced into the cortilymph space and, in reverse, cortilymph may be forced out via the intercellular spaces at the supporting cells and the basilar membrane. The evidence of spotty microsaturation of HRP in the basilar membrane in the present study can be viewed as possible evidence that the fluid transport in the basilar membrane is unidirectional: tracers are moved out from the corti-

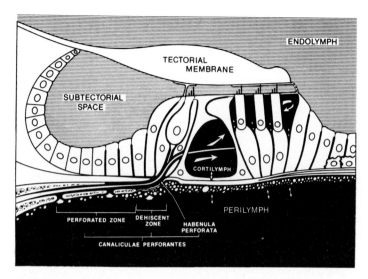

Fig. 25. An artist's conception of the perilymphatic communication routes between perilymph and cortilymph, depicting the canaliculi perforantes as the primary route and the transbasilar membrane route as the secondary route. The zona perforata would also facilitate direct contact of perilymph with the blood vessels in the spiral lamina (not shown). (Drawing by *Nancy Sally*.)

lymph space rather than entering into the cortilymph, although bidirectionality is also possibility. If the vessels that took up the tracers are indeed the venous type, then it may be that vessels near the basilar membrane are venous type to facilitate the removal of CO_2 and waste products diffused out from cortilymph through the basilar membrane.

In conclusion, although it is highly speculative, the results of the present morphological and tracer studies suggest that the concept of perilymphatic channels (canaliculae perforantes) proposed by *Schuknecht* [26] is valid. These channels may facilitate fluid passage between cortilymph and perilymph by rapid diffusion through the perineural spaces of the habenula perforata that serve as a major fluid passage system, as illustrated in figure 25. A secondary fluid passage (or reverse passage) is through the intercellular spaces of the pillar cells and, to a lesser extent, Deiters' cells, Hensen's cells and Claudius' cells from the basilar membrane. It is possible that this may be a unidirectional fluid flow from the cortilymph to the perilymph of the scala tympani, although a bidirectional flow may also exist. In addition, the bony pores may also facilitate fluid passage to the ganglion cells and to the CSF, as shown by *Sando* et al. [24].

Acknowledgment

The authors wish to thank Drs. *Harold F. Schuknecht*, *Robert S. Kimura* and *Dennis R. Trune* for their critical reviews of the manuscript and Dr. *Raúl Hinojosa* of The University of Chicago for generously providing the dialized ferritin used in the study. We are greatly endebted to *Ilija Karanfilov*, *Nancy Sally*, *James E. Douthitt*, and *Katherine Adamson* for their invaluable assistance.

This work was supported part by US Air Force contract No. F33615-80-C-0506 through the Aerospace Medical Research Laboratory, Wright-Patterson Air Force Base, Ohio, and a grant from The Deafness Research Foundation.

References

1 Angelborg, D.; Engström, H.: The normal organ of Corti; in Møller, Basic mechanisms in hearing, pp. 125–183 (Academic Press, New York 1973).

2 Asher, D.L.; Sando, I.: Perilymphatic communication routes in the auditory and vestibular system. Otolaryngol. Head Neck Surg. *89:* 822–830 (1981).

3 Berridge, J.B.; Oschman, J.L.: Transporting epithelia (Academic Press, New York 1972).

4 Cabezudo, L.M.: The ultrastructure of the basilar membrane in the cat. Acta oto-lar. *86:* 160–175 (1978).

5 De Lorenzo, A.J.D.; Shiroky, D.V.; Cohn, E.I.: Distribution of exogenous horseradish peroxidase in the perilymphatic and endolymphatic spaces of the guinea pig cochlea; in De Lorenzo, Vascular disorders and hearing defects, pp. 185–203 (University Park Press, Baltimore 1973).

6 Duvall, A.J., III; Sutherland, D.R.: Cochlear transport of horseradish peroxidase. Ann. Otol. Rhinol. Lar. *81:* 705–713 (1972).

7 Engström, H.: The cortilymph, the third lymph of the inner ear. Acta morph. neerl. scand. *3:* 195 (1960).

8 Engström, H.; Wersäll, J.: Is there a special nutritive cellular system around the hair cells in the organ of Corti? Ann. Otol. Rhinol. Lar. *62:* 508 (1952).

9 Hinojosa, R.: Transport of ferritin across Reissner's membrane. Acta oto-lar., suppl. 292 (1971).

10 Ilberg, C. von: Elektronenmikroskopische Untersuchungen über Diffusion und Resorption von Thoriumdioxyd an der Meerschweinchenschnecke. Arch. klin. exp. Ohr.-, Nas.-, Kehlk. Heilk. *192:* 384–400 (1968).

11 Ilberg, C. von; Vosteen, K.H.: Permeability of the inner ear membranes. Acta Otolaryngol. *67:* 165–170 (1960).

12 Jahnke, K.: Verteilung intrathekal applizierter Peroxydase in der Meerschweinchen-Cochlea. Arch. klin. exp. Ohr.- Nas.- KehlkHeilk. *202:* 418–422 (1972).

13 Karnovsky, M.J.: The ultrastructural basis of capillary permeability studies with peroxidase as a tracer. J. Cell Biol. *35:* 213–236 (1967).

14 Lawrence, M.; Nuttall, A.L.: Oxygen availability in tunnel of Corti measured by microelectrode. J. acoust. Soc. Am. *52:* 566–573 (1972).

15 Lim, D.J.: Surface ultrastructure of the cochlear perilymphatic space. J. Lar. Otol. *84:* 413–428 (1970).

16 Masuda, Y.; Sando, I.; Hemenway, W.G.: Perilymphatic communication routes in guinea pig cochlea. Archs Otolar. *94:* 240–245 (1971).

17 Misrahy, G.A.; De Jonge, B.R.; Shinabarger, E.W.; Arnold, J.E.: Effects of localized hypoxia on the electrophysiological activity of cochlea of the guinea pig. J. acoust. Soc. Am. *30:* 705–709 (1958).

18 Misrahy, G.A.; Hildreth, K.M.; Shinabarger, E.W.; Clark, L.C.; Rice, E.A.: Endolymphatic oxygen tension in the cochlea of the guinea pig. J. acoust. Soc. Am. *30:* 247–250 (1958).

19 Nadol, J.B., Jr.: Intercellular junctions in the organ of Corti. Ann. Otol. Rhinol. Lar. *87:* 70–81 (1978).

20 Nadol, J.B., Jr.: Intercellular fluid pathways in the organ of Corti of cat and man. Ann. Otol. Rhinol. Lar. *88:* 2–11 (1979).

21 Nomura, Y.: A histochemical study of a perilymph-cortilymph connection. Pract. Otol., Kyoto *61:* 469–474 (1968).

22 Rauch, S.: Entstehung, Transport und Resorption der Innenohrflüssigkeiten. Biochemie des Hörorgans: Einführung in Methoden und Ergebnisse, pp.278–315 (Thieme, Stuttgart 1964).

23 Ross, M.D.; Nuttal, A.L.; Wright, C.G.: Horseradish peroxidase acute ototoxicity and the uptake and movement of the peroxidase in the auditory system of the guinea pig. Acta oto-lar. *84:* 187–201 (1977).

24 Sando, I.; Wood, R.P., II; Masuda, Y.; Hemenway, W.G.: Perilymphatic communication routes in guinea pig cochlea. Ann. Otol. Rhinol. Lar. *80:* 826–834 (1971).

25 Schnieder, E.-A.: A contribution to the physiology of the perilymph. I. The origins of perilymph. Ann. Otol. Rhinol. Lar. *83:* 76–83 (1974).

26 Schuknecht, H.F.: Discussion of anatomy and physiology of peripheral auditory mechanisms; in Rasmussen, Windle, Neural mechanisms of the auditory and vestibular systems, pp.94–95 (Thomas, Springfield 1960).

27 Schuknecht, H.F.: Pathophysiology of the fluid systems of the inner ear; in Neff, Contributions to sensory physiology, vol.4, pp.75–93 (Academic Press, New York 1970).

28 Schuknecht, H.F.; Seifi, A.E.: Experimental observations on the fluid physiology of the inner ear. Ann. Otol. Rhinol. Lar. *72:* 687–712 (1963).

29 Tanaka, K.: Frozen resin cracking method for scanning electron microscopy of biological materials. Naturwissenschaften *59:* 77 (1972).

30 Tanaka, T.; Kosaka, N.; Takiguchi, T.; Aoki, T.; Takahara, S.: Observation on the cochlea with SEM. Scanning Electron Microscopy 1973, pp.428–433 (IIT Research Institute, Chicago 1973).

31 Tasaki, I.; Fernández, C.: Modification of cochlear microphonics and action potentials by KCl solution and by direct currents. J. Neurophysiol. *15:* 497–512 (1952).

32 Tonndorf, J.; Duvall, A.J., III; Reneau, J.P.: Permeability of intracochlear membranes to various vital stains. Ann. Otol. Rhinol. Lar. *71;* 801–841 (1962).

D.J. Lim, MD, Otological Research Laboratories, Department of Otolaryngology, Ohio State University College of Medicine, Columbus, OH 43210 (USA)

Adv. Oto-Rhino-Laryng., vol. 31, pp. 118–134 (Karger, Basel 1983)

Synaptic Structures of the Human Vestibular Ganglion [1]

Ken Kitamura, Robert S. Kimura [2]

National Medical Center Hospital, Tokyo, Japan, and Harvard Medical School and Massachusetts Eye and Ear Infirmary, Boston, Mass., USA

Introduction

The first study on the ultrastructure of the vestibular ganglion cells demonstrated bipolar ganglion cells in the goldfish [1]. Most of the cells were found to be myelinated and very few were unmyclinated. Several other authors reported similar ultrastructural findings in the vestibular ganglion of the rat, guinea pig, frog and cat [2–5]. It has been suggested that some of the vestibular ganglion cells in various species are multipolar [6], and recently it was demonstrated in the cat and the human by light microscopy [7,8]. The synaptic structure has also been displayed in the vestibular ganglion of the mouse [9].

There are very few studies, however, that deal with the fine structure of the human vestibular ganglion; furthermore, it is not clear whether there is a synaptic structure in the human vestibular ganglion. The purpose of this study is to describe the synaptic structures of the multipolar vestibular ganglion cells in the human.

Material and Methods

The inner ears of a 53-year-old male who died of congestive heart disease were fixed 4 h after death. He was reported to have no auditory or vestibular complaints. The temporal bones were fixed in 4% paraformaldehyde and 5%

[1] This work was supported by NINCDS Grant 5 R01 NSO3932-20.

[2] The authors wish to express their appreciation to Ms. *Carol Ota* for her help in the preparation of this manuscript for publication.

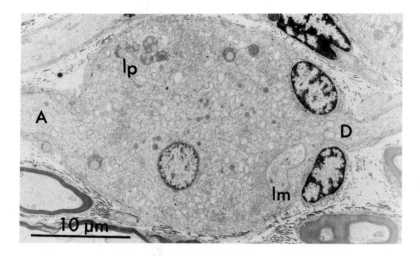

Fig. 1. Unmyelinated bipolar cell with a dendrite (D) and axon (A). Many mitochondria and endoplasmic reticulum are seen in the cytoplasm. Lipofuscin granules (lp) and lamellar membranous structures (lm) are also observed.

glutaraldehyde. The vestibular nerves were removed intact and postfixed in 1% phosphate-buffered osmium tetroxide. The tissue was stained en bloc with aqueous uranyl acetate, dehydrated in graded alcohols, exchanged with propylene oxide, and embedded in Epon. Serial sections of the right superior vestibular nerve were cut with an LKB ultratome, stained with lead citrate and examined with a Siemens Elmiskop 1 and Hitachi H 500.

Findings

The majority of the human vestibular ganglion cells are unmyelinated and invested by a thin cytoplasmic layer of one to three satellite cells. The size of the unmyelinated ganglion cells varies considerably from 18 to 50 μm. Their cell shapes are round to ellipsoidal. There are a few myelinated ganglion cells which show a rather uniform size of 36–46 μm in diameter. The nucleus is centrally or eccentrically located and contains one nucleolus in both myelinated and unmyelinated cells.

The vestibular ganglion cells are usually bipolar (fig. 1) as previously reported [1, 2]. The peripheral process of the bipolar cells is always thinner than the central process and the two processes are located at opposite poles

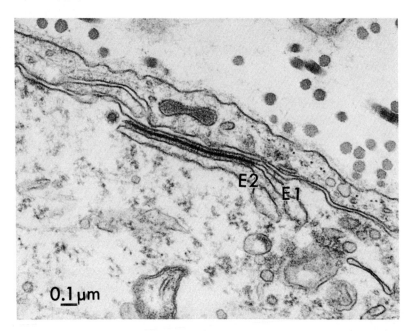

Fig. 2. Peculiar type of subsurface cistern in the unmyelinated cell. It is composed of two endoplasmic reticula (E1, E2). Note the two fused membranes of the endoplasmic reticulum (E1) adjacent to the plasmalemma.

of the perikarya. Their cytoplasms are rich in endoplasmic reticulum and contain numerous mitochondria. Neurofilaments appear prominent at the axon hillocks.

The ganglion cells also exhibit inclusion bodies of several types. There are numerous lysosomes and lipofuscin granules as well as lamellar membranous structures (fig. 1). Peculiar types of subsurface cisternae are encountered in the unmyelinated cells (fig. 2). Two membranes of the endoplasmic reticulum fuse to make a single dense line against the neuronal plasmalemma and the second endoplasmic reticulum is juxtaposed on the dense line. The deep surface of the second endoplasmic reticulum is studded sometimes with ribosome particles.

Fig. 3. a Unmyelinated multipolar cell with a dendrite (D1) directed peripherally. *b* Serial sections show two dendrites (D2, D3) in the bracketed area (**) of figure 3a. Their direction is almost perpendicular to the long axis of the cell. *c* An axon (A) directed centrally is observed in the bracketed area (*; fig. 3a) from serial sections of the same multipolar cell.

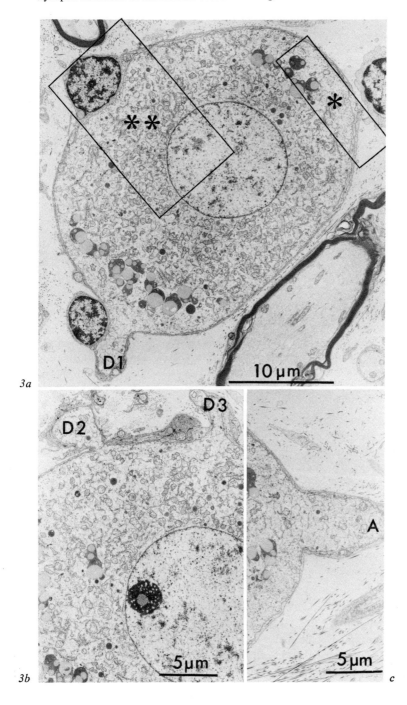

3a

3b   *c*

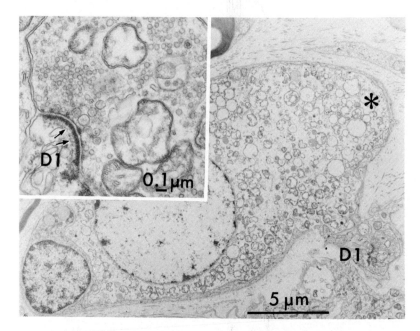

Fig.4. Unmyelinated multipolar cell with synapse. Serial sections demonstrate the synapse on a small dendritic process (D1) and a second dendrite (D2; fig. 5) that appears in the right upper pole of the cell (*). A high magnification showing many agranular spherical vesicles in the nerve terminal and fine dense bodies (arrows) aligned parallel to the thick postsynaptic membrane in the dendrite (insert).

In addition to the bipolar cells, multipolar cells are observed in our study (fig. 3). There are large and small multipolar cells. The large ones are almost equal in size to the bipolar cells. The neuronal cytoplasms of the multipolar cells exhibit less free ribosomes and less endoplasmic reticulum. However, these characteristics are not adequate to classify bipolar cells as dark cells and multipolar cells as light cells, as some propose. Some of the bipolar cells are demonstrated to contain fewer cytoplasmic structures than the cell shown in figure 1. Nerve synapses have not yet been observed on these large multipolar cells or on their cell processes near the perikarya.

Nerve fiber synapses are observed on the perikarya and/or the dendrites of the small multipolar cells. The morphological characteristics of nerve terminals appear to be two different types. The first type is shown in figures 4–7. The ganglion cell in figure 4 is spindle-shaped and it has a long axis of 24 µm. The cytoplasm of the cell is characterized by the inclusion of many

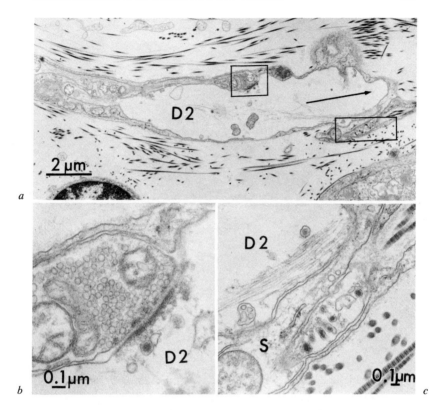

Fig. 5. a Dendrite (D2) of the cell shown in figure 4 followed in serial sections. It comes out from the peripheral pole of the cell heading toward the end organ (arrow). High magnifications of the two bracketed areas demonstrate the different types of nerve fibers. *b* Nerve terminal on the left in figure 5a makes a synapse on the dendrite. Most of the synaptic vesicles are spherical. *c* No synapse is observed on the nerve terminal on the right containing elongated vesicles. S = Schwann's cell.

dense-cored vesicles (110–120 nm). Two processes come out from the opposite poles and a third small one emerges close to the peripheral process. A synapse is found on the peripheral dendritic process at about a distance of 20 μm from the perikaryon and also on a smaller dendrite (fig. 4, 5). The juxtaposed membranes between nerve terminals and dendrites of ganglion cells show local differentiations characterized by an increased density. These specialized zones are about 0.8 μm in length and a synaptic cleft is about 30 nm. Small dense bodies are lined up parallel to the thickened postsynaptic membrane in the dendrite. The vesicles observed in these nerve terminals

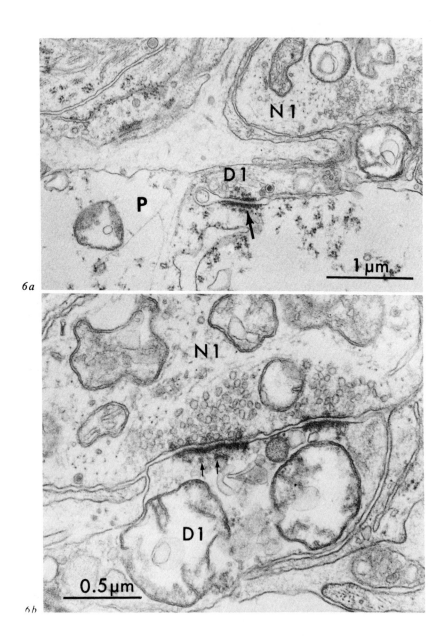

6a

6b

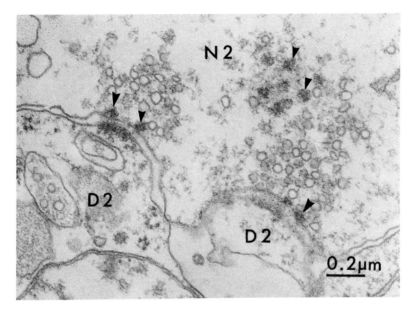

Fig. 7. Synapses between a nerve terminal (N2) and a second small dendritic process (D2) of the same multipolar cell as shown in figure 6. Presence of the small dense bodies similar to the presynaptic bodies (arrow heads) in the deeper zone suggests third synaptic contact to the dendrite. Most of the synaptic vesicles are spherical.

sometimes accumulate at the cone-shaped patches of dense material. Most of the vesicles are agranular spherical (40–50 nm in diameter), and a small number of them are granular spherical vesicles (80–120 nm in diameter, fig. 4, 5). Small unmyelinated nerve fibers containing vesicles are often found adjacent to the perikaryon or the dendrites of the ganglion cell (fig. 5). Most of the vesicles in these nerve fibers are granular elongated (137–170 nm along the long axis) and a few of them are agranular elongated (68–85 nm along the long axis).

Fig. 6. a Nerve terminal (N1) abuts a small dendritic process (D1) of the unmyelinated multipolar cell. Note the multivesicular body and the dense-cored vesicle in the dendritic process. Desmosomal contact (arrow) is observed between the perikaryon (P) and its dendritic process. *b* High magnification of the synapse from the serial sections of the same cell shown in figure 6a. Many spherical vesicles are observed in the nerve terminal (N1) and some of them are clustered at the thickened presynaptic membrane. Small dense bodies (arrows) are lined up against the postsynaptic thickened membrane.

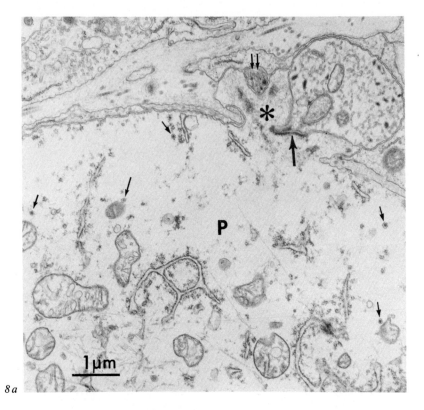

8a

Fig.8. a An unmyelinated ganglion cell (P) showing a finger-like cytoplasmic extension (*) which makes a synaptic contact (large arrow). A profile of another small nerve ending (two arrows) is shown in an adjacent area. Note many dense-cored vesicles in the perikaryon (arrows). *b* High magnification of the synapse from the serial sections of the same area shown in figure 8a. Note a finger-like cytoplasmic extension (*) and numerous granular and agranular elongated vesicles in the nerve terminal.

Similar synapses are found on the dendritic processes of another small unmyelinated multipolar neuron. The cell is ellipsoidal in profile and its size is 25 μm. It shows the central and peripheral processes coming out of opposite sides and two short dendritic processes. One of the short dendritic processes bends just after arising from the perikaryon and forms a synapse with the nerve terminal (fig. 6). The small dendritic process contains a multivesicular body and dense-cored vesicle. Desmosomal contact is observed between the perikaryon and its dendritic process. Specialization of the membranes between the nerve terminal and the dendritic process is similar to

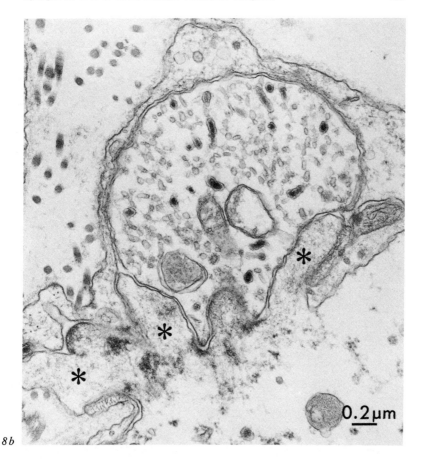

8b

those shown in figure 4. The synaptic cleft is relatively narrow (17 nm). The nerve terminal contains many agranular spherical vesicles (about 35 nm in diameter). A synaptic structure is also observed on the second small dendritic process (fig. 7). The process also contains a multivesicular body and dense-cored vesicle. The nerve terminal contains small bodies similar to the dense patches on the presynaptic membrane. These dense bodies may establish another synapse on the same dendrite. Most of the vesicles in the nerve terminal are agranular spherical (40–55 nm in diameter).

A second type of synapse is demonstrated on the dendrite and peri-karyon of the unmyelinated neurons (fig. 8, 9). The ganglion cell in figure 8 is almost circular and its size is 22 μm. Many dense-cored vesicles are observed in the perikaryon. Two cell processes come out of opposite sides but are

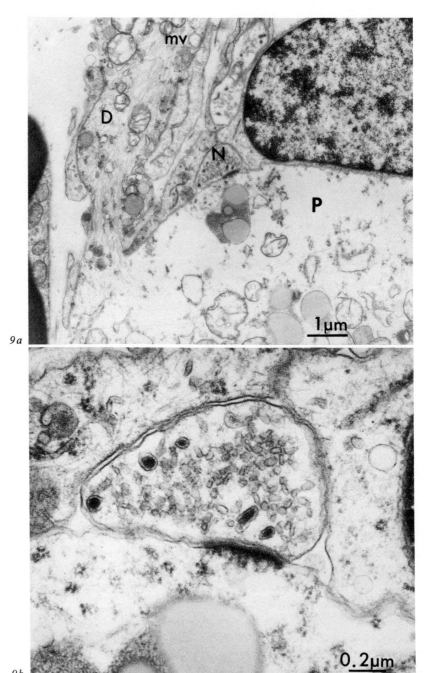

9a

9b

oriented at right angles to the central-peripheral axis. On the third process, a finger-like cytoplasmic extension, a nerve terminal, establishes synaptic contact (fig. 8). The nerve terminal contains many granular elongated vesicles having a long axis of 84–250 nm. The agranular vesicles are numerous and their shapes vary from round to flattened (63–105 nm in diameter). The morphological specializations of the synaptic contact between the nerve terminal and the dendrite are similar to those seen in the previous synaptic structures.

A nerve terminal similar to the second type is found making a synaptic contact directly on the perikaryon of another neuron (fig. 9). The ganglion cell is oval and 18 μm in diameter. The cytoplasm includes many dense-cored vesicles. Multivesicular bodies are observed in the large process which is pointed toward the peripheral direction. The other two processes are small and emerge close to the large one. There is a remarkable thickening of the perikaryal plasmalemma at the nerve contact area, and the nerve fiber contains many agranular (long axis 48–95 nm) as well as granular flattened vesicles (long axis 95–140 nm).

Discussion

The presence of multipolar vestibular ganglion cells was briefly described in the guinea pig and demonstrated in the cat [3, 7]. In the human vestibular ganglion, a multipolar cell was also reported [8]. The perikaryon of the dark cell of the mouse vestibular ganglion was found to exhibit a synaptic contact with a nerve fiber [9]. However, there has been no study to demonstrate the synaptic structures in relation to the multipolar neurons in the human vestibular ganglion.

In the present study two types of multipolar cells are observed in the human vestibular ganglion. One of them is large in size, similar to the bipolar cells, and no synapse is found. The other one is smaller and has nerve fiber synapses. Synaptic terminals are further classified into two types (fig. 10). The ultrastructural features of ganglion cells which receive synaptic contacts can

---

Fig. 9. a Nerve terminal (n) making a synaptic contact on the perikaryon (P) of the multipolar cell. Numerous neurofilaments and multivesicular bodies (mv) are observed in the dendrite (D). *b* High magnification of the same synapse showing prominent postsynaptic thickening of the plasmalemma. Note the numerous agranular flattened vesicles as well as granular flattened vesicles in the nerve terminal.

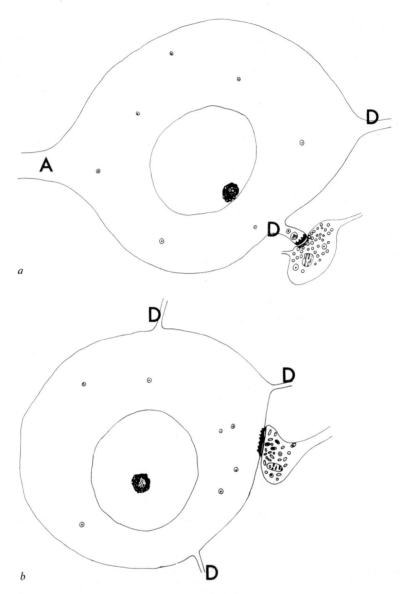

Fig. 10. Schematic drawing of two types of synaptic terminals observed in the human vestibular ganglion. Morphological specializations at the synaptic contact are similar in both types of synapse. The first terminal in figure 10a is demonstrated on the dendrite and spherical vesicles are in the majority. On the other hand, the second type in figure 10b is observed on the perikaryon or on the dendrite and ellipsoidal vesicles are numerous. A = Axon; D = dendrite.

not be fully determined because of postmortem changes but there is no obvious difference between the perikarya of ganglion cells which have two different types of synaptic structures.

The most significant differences between the two types of nerve terminals are the locations and shapes of synaptic vesicles. The first type of synaptic terminal is demonstrated on the process, and spherical vesicles are in the majority. The second type is exhibited on the perikaryon and the dendrite, and ellipsoidal vesicles are numerous. Morphological specializations at the synaptic contact are, in principle, remarkably similar throughout most of the vertebrate nervous system. However, some variations of type 1 and 2 synapses have been proposed [10]. The first type of synapse in the present study has morphological characteristics similar to type 1 synapses which occur on dendritic spines and on the shafts of smaller dendritic branches with pronounced accumulation of dense material on the cytoplasmic face of the postsynaptic membrane.

The second type of synapse in our study occurs on the perikaryon in a manner similar to the type 2 synapse described by *Gray* [10]. However, it has prominent postsynaptic thickenings which are not morphologically characteristic of Gray's type 2 synapses. The nerve terminal corresponding to the second type is observed among the geniculate ganglion cells [11].

It is well-known that synaptic vesicles are osmotically sensitive and change their shapes both during and after fixation with aldehyde [12, 13]. In the present case, however, one would expect a uniform effect to be observed in all the synaptic structures of the same specimen, whereas the shape of synaptic vesicles differs even in the same thin section. This would suggest less probability of artifactual influence on the different shapes of synaptic vesicles. The morphological differences in synaptic vesicles have been attributed functional significance by several authors [14, 15]. They suggest that the spherical vesicles are concerned with excitation and the ellipsoidal vesicles with inhibition. On the other hand, *Rastad* [16, 17] reported that the average length of the synaptic vesicles is significantly longer in the axo-somatic than in the axo-dendritic terminals of spinocervical tract cells, and that length/width ratios for the synaptic vesicles are not significantly different between the axon terminals of the inhibitory spinal interneurons and the excitatory spinocervical tract cells.

Although we did not attempt to trace the origin of the nerve fibers which make synaptic contacts on the vestibular ganglion cells in the present study, three possible origins could be suggested. They may be axons from 'receptor-receptor' fibers or the vestibular efferent system or autonomic fibers.

The existence of interrelations between vestibular receptors was demonstrated outside of the brain stem by an electrophysiological method [18]. They suggested the presence of 'receptor-receptor' fibers which projected onto the crista ampullaris of the superior semicircular canal from several other ipsilateral vestibular receptors, especially the crista of the horizontal semicircular canal. In their more recent study they showed an inhibitory feedback pathway arising from the horizontal canal crista and going back to the horizontal canal crista, or from the vertical canal crista and back to the vertical canal crista [19]. It has been proposed that the multipolar ganglion cell is responsible for these interreceptor activites [7]. Our morphological data allows one to speculate that the neuron establishes synaptic contact with a branch of the afferent fiber and sends an inhibitory feedback fiber to the same or different receptor organ.

The course of the vestibular efferent pathway was originally described by *Gacek* [20]. Findings from physiological and morphological studies indicate that the vestibular efferent neurons from the brain stem often send collaterals to innervate several cristae in one labyrinth and, less frequently, fibers to the contralateral labyrinth [21,22]. In addition, antidromic stimulation of the efferent pathway produced longer latency synaptic responses [21]. The route of these synaptic responses remains unknown and it cannot exclude the possible influence of the vestibular afferent neurons on the efferent pathway in the brain stem. It is, however, suggested that the vestibular efferent pathway has a direct contact on the vestibular ganglion cells or it has a post-ganglionic neuron at the level of the vestibular ganglion.

It is well-documented by fluorescent and electron microscopic studies [23,24] that there are adrenergic blood vessel-independent nerve fibers in the vestibular ganglion of the cat, rabbit and human. Adrenergic nerve terminals show many small granular vesicles and a variable number of large granular vesicles [25]. This kind of nerve terminal is not demonstrated in our study; however, the possibility of structural change by fixatives cannot be ruled out [26]. Parasympathetic efferent fibers are also reported to supply fibers to the vestibular nerve and it is proposed that they have synaptic contacts on the vestibular ganglion [9,27]. Synaptic structures were observed on the small spiral ganglion cells in the human and monkey by *Kimura and Ota* [28] who speculated that these neurons were either sensory or parasympathetic in origin.

The first type of synaptic terminal in our study contains predominantly small agranular vesicles and a small number of large granular vesicles which show similar morphologic characteristics to the terminals of cholinergic autonomic nerves [25,26]. The ultrastructural features of the vestibular ganglion cells which have nerve fiber synapses cannot be fully evaluated because of their postmortem changes but dense-cored vesicles are numerous in their cytoplasm. It is generally accepted that monoamines are stored in dense-cored vesicles and they have been demonstrated in both the sympathetic and parasympathetic ganglion cells [29,30]. Thus, one cannot rule out the possibility that these ganglion cells are parasympathetic ganglion cells and dual autonomic innervation on the vestibular ganglion cell in accordance with two types of synaptic structures.

References

1   Rosenbluth, J.; Palay, S.L.: The fine structure of nerve cell bodies and their myelin sheaths in the eighth nerve ganglion of the goldfish. J. biophys. biochem. Cytol. *9:* 853–877 (1961).

2   Rosenbluth, J.: The fine structure of acoustic ganglia in the rat. J. Cell Biol. *12:* 329–359 (1962).

3   Ballantyne, J.; Engström, H.: Morphology of the vestibular ganglion cells. J. Lar.Otol. *83:* 19–42 (1969).

4   Kuleshova, T.F.: On neuronal structure of the vestibular ganglion in the frog. Arkh. Anat. Gistol. Embriol. *74:* 72–76 (1978).

5   Richter, E.; Spoendlin, H.: Scarpa's ganglion in the cat. Acta oto-lar. *92:* 423–431 (1981).

6   Bovero, A.: Connessioni simpatiche del ganglio vestibolare del nervo acustico. Arch. Ital. Otolaryngol. *25:* 41–45 (1914).

7   Chat, M.; Sans, A.: Multipolar neurons in the cat vestibular ganglion. Neuroscience *4:* 651–657 (1979).

8   Ylikoski, J.; Belal, A., Jr.: Human vestibular nerve morphology after labyrinthectomy. Am. J. Otolaryng. *2:* 81–93 (1981).

9   Ehrenbrand, F. von; Wittemann, G.: Über synaptische Strukturen im Ganglion vestibulare der Maus. Anat. Anz. *126:* 300–308 (1970).

10 Gray, E.G.: Axo-somatic and axo-dendritic synapses of the cerebral cortex: an electron microscope study. J. Anat. *93:* 420–433 (1959).

11 Kitamura, K.; Kimura, R.S.; Schuknecht, H.F.: The ultrastructure of the geniculate ganglion. Acta oto-lar. *93:* 175–186 (1982).

12 Valdivia, O.: Methods of fixation and the morphology of synaptic vesicles. J. comp. Neurol. *142:* 257–274 (1971).

13 Nakajima, Y.: Fine structure of the synaptic endings on the Mauthner cell of the goldfish. J. comp. Neurol. *156:* 375–402 (1974).

14 Uchizono, K.: Characteristics of excitatory and inhibitory synapses in the central nervous system of the cat. Nature, Lond. *207:* 642–643 (1965).

15 Nadol, J.B., Jr.; De Lorenzo, A.J.D.: Observations on the abdominal stretch receptor and the fine structure of associated axo-dendritic synapses and neuromuscular junctions in Homarus. J. comp. Neurol. *132:* 419–444 (1968).

16 Rastad, J.: Morphology of synaptic vesicles in axo-dendritic and axo-somatic col-
    lateral terminals of two feline spinocervical tract cells stained intracellularly with
    horseradish peroxidase. Exp. Brain Res. *41:* 390–398 (1981).
17 Rastad, J.: Ultrastructural morphology of axon terminals of an inhibitory spinal
    interneuron in the cat. Brain Res. *223:* 397–401 (1981).
18 Gribenski, A.; Caston, J.: Fibers projecting onto the crista ampullaris of the vertical
    anterior semicircular canal from other ipsilateral vestibular receptors in the frog
    (Rana esculenta). Pflügers Arch. *349:* 257–265 (1974).
19 Caston, J.; Gribenski, A.: Innervation of a vestibular receptor in the frog: existence
    of a feedback loop? J. comp. Physiol. *133:* 63–69 (1979).
20 Gacek, R.R.: Efferent component of the vestibular nerve; in Rasmussen, Windle,
    Neural mechanisms of the auditory and vestibular systems, pp.276–284 (Thomas,
    Springfield 1960).
21 Khalsa, S.B.S.; Schwarz, D.W.F.; Fredrickson, J.M.; Landolt, J.P.: Efferent
    vestibular neurons. Electrophysiological evidence for axon collateralization to cristae
    ampullares in the pigeon *(Columba livia)*. Acta oto-lar. *92:* 83–88 (1981).
22 Schwarz, I.E.; Schwarz, D.W.F.: Fredrickson, J.M.; Landolt, J.P.: Efferent vestibu-
    lar neurons. A study employing retrograde tracer methods in the pigeon *(Columba
    livia)*. J. comp. Neurol. *196:* 1–12 (1981).
23 Densert, O.: A fluorescence and electron microscopic study of the adrenergic inner-
    vation in the vestibular ganglion and sensory areas. Acta oto-lar. *79:* 96–107 (1975).
24 Ylikoski, J.; Partanen, S.; Palva, T.: Adrenergic innervation of the eighth nerve and
    vestibular end organs in man. Arch. Oto-Rhino-Lar. *224:* 17–23 (1979).
25 Burnstock, G.: The ultrastructure of autonomic cholinergic nerves and junctions; in
    Tuček, Progress in brain research: the cholinergic synapse, vol.49 (Amsterdam 1979).
26 Eränkö, O.; Klinge, E.; Sjöstrand, N.O.: Different types of synaptic vesicles in axons
    of the retractor penis muscle of the bull. Experientia *32:* 1335–1337 (1976).
27 Ross, M.D.: The general visceral efferent component of the eighth cranial nerve.
    J. comp. Neurol. *135:* 453–478 (1969).
28 Kimura, R.S.; Ota, C.Y.: Nerve fiber synapses on primate spiral ganglion; in Lim,
    Abstr. of the Mid-Winter Meet. of the Association for Research in Otolaryngology,
    Columbus 1981, p.82.
29 Dixon, J.S.: The fine structure of parasympathetic nerve cells in the otic ganglia of
    the rabbit. Anat. Rec. *156:* 239–252 (1966).
30 Hökfelt, T.: Distribution of noradrenaline storing particles in peripheral adrenergic
    neurons as revealed by electron microscopy. Acta phys. scand. *76:* 427–440 (1969).

R.S.Kimura, PhD, Department of Otolaryngology, Massachusetts Eye and Ear
Infirmary, 243 Charles Street, Boston, MA 02114 (USA)

Adv. Oto-Rhino-Laryng., vol. 31, pp. 135–147 (Karger, Basel 1983)

Caprine Beta Mannosidosis: Temporal Bone Pathology

J.T. Benitez, G.E. Lynn, M.Z. Jones

William Beaumont Hospital, Division of Otoneurology, Royal Oak, Mich.;
Wayne State University, Departments of Audiology and Neurology, Detroit, Mich.;
Michigan State University, Department of Pathology, East Lansing, Mich., USA

Introduction

Inherited disorders of glycoprotein catabolism are associated with marked neurological dysfunction in man and other species. α-mannosidosis, for example, is characterized by neurological deficits, storage and excretion of oligosaccharides with terminal α-linked mannose residues and deficiency of α-*D*-mannosidase activity. The first human case was described by *Öckerman* [7] in 1967. Since then, several other cases have been reported. Clinical features have included a dysmorphic gargoyle-like facial appearance resembling Hurler syndrome, moderate to severe mental retardation and lack of motor coordination. Some cases presented recurrent infections, lens opacification, hearing loss and hepatosplenomegaly. Hearing impairment was detected in 5 patients of 14 cases reported by *Kistler* et al. [5].

In 1975, *Jones* et al. [2], identified in Michigan a new caprine neurovisceral storage disease as β-mannosidosis. The clinical features of this disease were observed in Australia in 1973 by *Hartley and Blakemore* [1]. *Jones and Dawson* [3], *Jones and Laine* [4] and *Matsuura* et al. [6] have described the clinical, pathological, genetic and biochemical characteristics. The disease is associated with the deficiency of β-*D*-mannosidase and accumulation of tissue oligosaccharides. The human counterpart has not been identified as yet. It is inherited as an autosomal recessive trait. The clinical features include dome-shaped skull, narrow palpebral fissures, elongated muzzle, hyperextension of pastern joints, contraction of carpal joints, intention tremor, pendular nystagmus and deafness. The intention tremor and the nystagmus abate after about 3 weeks of age.

Microscopic examination revealed fine to coarse vacuolation in neurons and other cell types; extensive deficiency of central nervous system myelination with axonal lesions. Histopathological studies of the temporal bones of a 4-week-old doe affected with β-mannosidosis and its twin, which was unaffected by the disease, are the basis for this report.

Material and Methods

About 7 h after the unexpected death during the night of the unaffected animal (V-39), 10% formalin solution was injected into the bulla of the temporal bones. The affected animal (V-38) was sacrificed on the same day by decapitation and the bullae were injected with 10% formalin solution. The temporal bones of the 2 goats were removed for histological study. Every fifth section was stained with hematoxylin and eosin and additional sections with periodic acid-Schiff (PAS).

Auditory Evoked Potential Findings. Short-latency auditory evoked potential recordings obtained within 24 h prior to death of each animal are shown in figure 1. Recordings obtained from V-39, the unaffected female goat (fig. 1) show five well-defined waves or components from stimulation of each ear when the earphone was delivering a click intensity of 128 dB pSPL. Wave I latency and interpeak latencies are in milliseconds. These recordings are very similar to those seen in other mammalian species. Recordings obtained from V-38, the affected female goat (fig. 1) were flat showing no measurable evoked electrical response from either ear when stimulating at 128 and 136 dB pSPL. These records were interpreted to indicate the presence of a marked abnormality of the peripheral auditory system bilaterally. Clinically, the animal showed no awareness to high-intensity environmental sounds consistent with deafness.

Fig. 1. a AEP tracings from V-39, the unaffected goat. Five well-defined waves are shown in the recordings from stimulation of each ear at 128 dB pSPL. Wave I latency and interpeak latencies are in milliseconds. *b* AEP tracings from V-38, the affected goat showing no measurable electrical activity from either ear with stimulus intensities of 128 and 136 dB pSPL. RE = Right ear; LE = left ear.

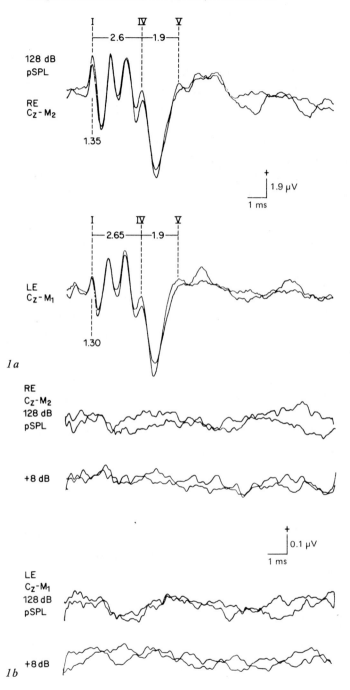

1a

1b

Temporal Bone Histopathology

Right Temporal Bone Affected Animal V-38. There is stenosis of the cartilaginous portion of the external auditory meatus. It is only about one third of the diameter of control animal V-39. The bony canal shows no abnormality. The tympanic membrane is normal in appearance with moderate thickening. The mucosa of the middle ear shows hypertrophy at the level of the oval and round windows with dense fibrosis and numerous histiocytes. The ossicular chain is normal. There is no fixation of the stapes.

In the inner ear, the cochlea (fig. 2) shows collapse of Reissner's membrane, more severe at the apical end. At this level, it is lying partially on the stria vascularis and remnants of the organ of Corti. In the middle and basal turns, the Reissner's collapse is only moderate. The tectorial membrane is missing throughout. In the middle and basal turns of the cochlea, the scala tympani shows moderate fibrosis with histiocytes. The population of neurons in the spiral ganglion is normal and they are normal in appearance. There are no abnormalities of the nerve fibers of the cochlear nerve. Histiocytes can be identified around the nerve fibers throughout the modiolus. Figure 2 shows the cochlea of control animal V-39 for comparison. There is no collapse of Reissner's membrane. The elements of the cochlear duct show moderate to severe postmortem changes. Hair cells in the organ of Corti can be identified. The tectorial membrane is present throughout. Neurons in the spiral ganglion show severe postmortem changes but the population is normal. The cochlear nerve is normal. Figure 3 shows the saccule of affected animal. The saccular wall is collapsed lying on the maccula. In one area, it is folded upon itself forming a double wall structure with histiocytes. The sensory epithe-

---

Fig. 2. a Right cochlea of affected animal V-38 showing dysplasia of the cochlear duct. Remnants of the organ of Corti can be identified only in the upper basal turn. There are areas of fibrosis with histiocytes (arrows). Spiral ganglion neuron population is normal. The cochlear nerve shows no abnormality. *b* Right cochlea of unaffected animal V-39. Reissner's membrane is in normal position. Structures in the cochlear duct and spiral ganglion neurons show only postmortem changes. Hematoxylin-eosin. × 22.

Fig. 3. a Right saccule of affected animal V-38. The saccular wall is collapsed upon the macula. In the posterior region it forms a double-walled structure with histiocytes. *b* Right saccule of unaffected animal V-39. There is no collapse of the saccular wall. The macula shows postmortem changes. Hematoxylin-eosin. × 42.

Fig. 4. a Right round window of affected animal V-38. The round window membrane (RM) is covered by dense fibrosis (F) with histiocytes. There is moderate body stenosis. *b* Right round window of unaffected animal V-39 showing thickening of the round window membrane (RM). There is no fibrosis covering it. Hematoxylin-eosin. × 18.

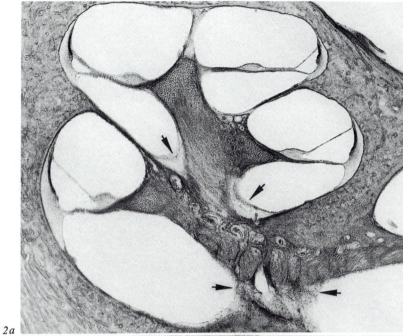

2a

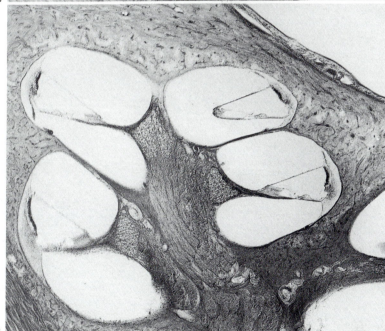

2b

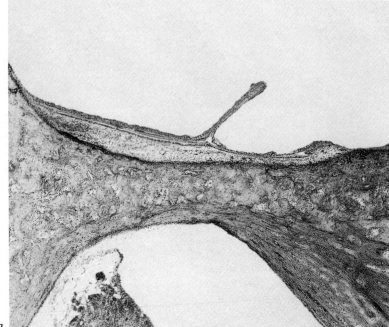

3a

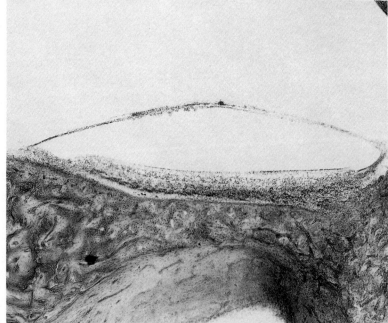

3b

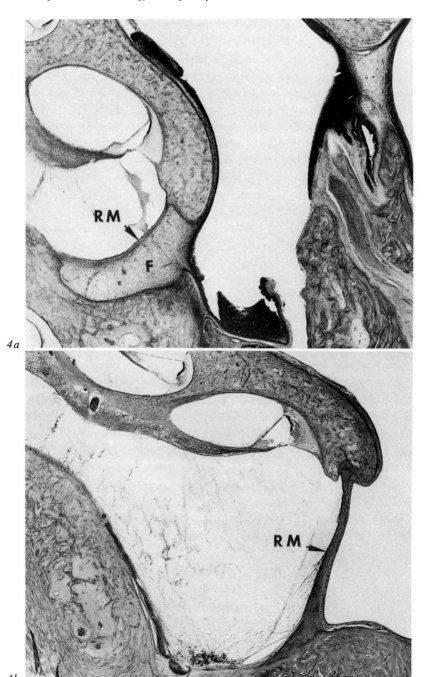

4a

4b

lium of the maccula shows only postmortem changes. Figure 3 shows the saccule in the unaffected animal without collapse of the saccular wall. The sensory epithelium of the macula shows postmortem changes. Figure 4 shows the round window of the affected animal. There is very dense fibrosis covering the round window membrane. Numerous histiocytes can be identified. Moderate bony stenosis is present in the affected but not the control animal. Figure 4 shows moderate thickening of the mucous membrane covering the round window membrane of the unaffected animal for comparison. The utriculocanalicular system is histologically normal. The vestibular nerve and Scarpa's ganglia are normal.

Left Temporal Bone Affected Animal V-38. Histological findings of the external and middle ear are similar to those of the right ear. The hypertrophy of the middle ear mucosa at the level of the oval and round windows is not as severe as in the right ear.

In the cochlea (fig. 5), Reissner's membrane is in an anatomical position. The elements of the cochlear duct show moderate to severe postmortem changes. In the organ of Corti, internal and external hair cells can be identified. The tectorial membrane is present throughout. There are areas of fibrosis in the middle turn of the cochlea with histiocytes. The neurons in the spiral ganglion show moderate postmortem changes with normal population. The nerve fibers of the cochlear nerve are normal. Histiocytes can be identified along the nerve fibers. Figure 5 shows the left cochlea of the control animal with similar histological findings except for the absence of histiocytes. The saccule shows the same histological appearance as the opposite ear (fig. 6). In the round window (fig. 7) of the affected animal, the round window membrane is covered by dense fibrosis with histiocytes. There is moderate

---

Fig. 5. a Left cochlea of affected animal V-38. There is no dysplasia of the cochlear duct; its structures show moderate to severe postmortem changes. Inner and outer hair cells can be identified throughout. The spiral ganglion and the cochlear nerve are normal. There are areas of fibrosis with numerous histiocytes (arrows). *b* Left cochlea of unaffected animal V-39 shows only postmortem changes. Hematoxylin-eosin. × 24.

Fig. 6. a Left saccule of affected animal V-38 showing collapse of the saccular wall; in the posterior region forms a double-walled structure with histiocytes. The macula shows postmortem changes. *b* Left saccule of unaffected animal. There is no collapse of the saccular wall. Hematoxylin-eosin. × 42.

Fig. 7. a Left round window of affected animal V-38. The round window membrane (RM) is covered by dense fibrosis (F) with histiocytes. There is moderate bony stenosis. *b* Left round window of unaffected animal V-39. There is no fibrosis covering the round window membrane (RM). Hematoxylin-eosin. × 18.

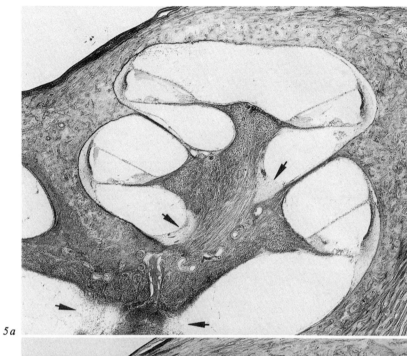

5a

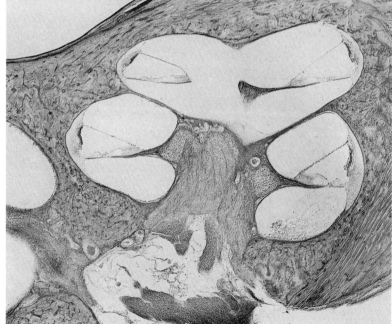

5b

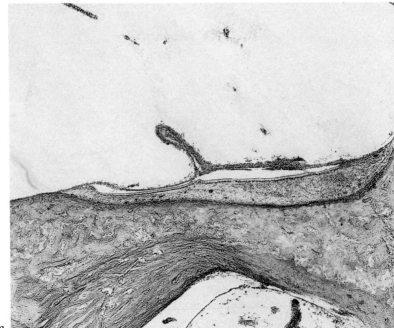

6a

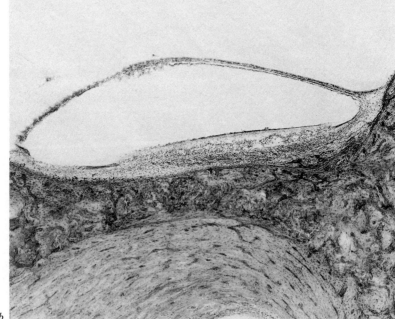

6b

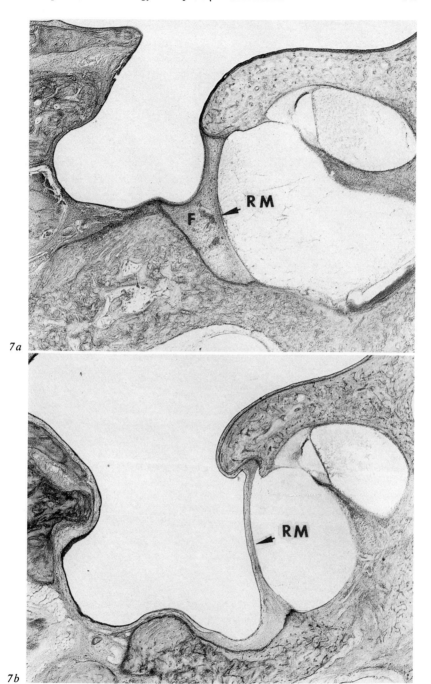

7a

7b

bony stenosis. Figure 7 shows the round window of the control ear for comparison. As in the opposite ear, the utriculocanalicular system is histologically normal. Neurons in the Scarpa's ganglia show postmortem changes with normal population. The vestibular nerve is normal.

Discussion

The histological study of the right ear of affected animal V-38 showed pathologic changes involving the cochlea and the saccule which is consistent with *Schiebe* [8] type inner ear dysplasia. In the left ear, structures in the cochlear duct were preserved despite the presence of a definite saccular lesion. It is possible that biochemical changes had taken place, which interfere with the electrical activity of the cochlea and, for this reason, no evoked potentials were obtained from stimulation of this ear.

The significance of the histiocytes found in the inner ear spaces and in the modiolus is not clear in this study. Are these macrophages transporters of storage material? Other investigators, using biochemical and ultrastructural studies, indicate that tissue storage of oligosaccharides is associated with lucent lyposomal storage vacuoles. The relationship between the inner ear lesions and the biochemical perturbations in β-mannosidosis remains to be determined.

Acknowledgements

We are indebted to *Doylene Lane-Szopo* and *Elizabeth Dilley* for histological preparations and Mr. *Arthur Bowden* for photomicrography.

This study was supported in part by Grant 78-4 from the William Beaumont Hospital Research Institute (J.T.B.) and Grant NS-16886 from the National Institutes of Health (M.Z.J.).

References

1   Hartley, W.J.; Blakemore, W.F.: Neurovisceral storage and dysmyelinogenesis in neonatal goats. Acta neuropath. *25:* 325–333 (1973).
2   Jones, M.Z.; Cunningham, J.G.; Dade, A.W.; et al.: Caprine neurovisceral storage disorder resembling mannosidosis. Soc. Neurosci. Abstr. *5:* 513 (1979).
3   Jones, M.Z.; Dawson, G.: Caprine β-mannosidosis: inherited deficiency of β-*D*-mannosidase. J. biol. Chem. *256:* 5185–5188 (1981).

4   Jones, M.Z.; Laine, R.A.: Caprine oligosaccharide storage disorder: accumulation of β-mannosyl(1-4)-β-*N*-acetylglucosaminyl(1-4)-β-*N*-Acetylglucosamine in brain. J. biol. Chem. *256:* 5181–5184 (1981).
5   Kistler, J.P.; Lott, I.T.; Kolodny, E.H.; et al.: Mannosidosis. Archs Neurol., Chicago *34:* 45–51 (1977).
6   Matsuura, F.; Laine, R.A.; Jones, M.Z.: Oligosaccharides accumulated in the kidney of a goat with β-mannosidosis: mass spectrometry of intact permethylated derivatives. Archs biochem. Biophys. *211:* 485–493 (1981).
7   Öckerman, P.A.: A generalized storage disorder resembling Hurler's syndrome. Lancet *ii:* 239–241 (1967).
8   Schiebe, A.: A case of deaf-mutism with auditory atrophy and anomalies of development in the membranous labyrinth of both ears. Archs Otolar. *11:* 12 (1892).

J.T. Benitez, MD, William Beaumont Hospital, Division of Otoneurology,
Royal Oak, MI 48072 (USA)

Adv. Oto-Rhino-Laryng., vol. 31, pp. 148–154 (Karger, Basel 1983)

Pathological Findings in the Cochlear Duct due to Endolymphatic Hemorrhage

T. Ishii, M. Toriyama, T. Takiguchi

Department of Otolaryngology, Tokyo Women's Medical College,
Medical Center of National Hospitals, Tokyo, and Hiroshima City Hospital,
Hiroshima, Japan

Introduction

Inner ear hemorrhage is found in cases with leukemia, leukemoid reaction of malignant neoplasma, DIC (Disseminated Intravascular Coagulation) syndrome or hemorrhagic diathesis of any causes. Blood clot is frequently found in the perilymphatic space. The authors experienced 2 patients who died of adenocarcinoma of the stomach who showed intense hemorrhagic diathesis before death. Before death, they developed sudden hearing loss in each ear. Two of their ears showed marked endolymphatic hemorrhage of the cochlea and one ear showed mainly perilymphatic bleeding. In the latter ear, the term of onset of deafness until death was 59 days. Blood was hardly seen in the cochlear duct, but several pathological changes showed residual findings of endolymphatic hemorrhage of the cochlea. The authors assume the blood in the cochlear duct disappeared after 2 months and the cochlear duct showed several characteristic findings.

Case Reports

Case 1
A 38-year-old man died of leukemoid reaction due to bone metastasis of gastric carcinoma. A few weeks before death, the platelet number became extremely low. 13 days prior to death, he suddenly developed total deafness in the left ear. The pure tone audiometry was performed and he did not respond to all sounds of 125–8,000 Hz on the left ear. The next day his right hearing was lost completely. He became completely deaf and did not recover from deafness before his death.

Temporal Bone Findings
Right Ear. A massive bleeding was observed both in the perilymphatic and endolymphatic spaces (fig. 1). Reissner's membrane was distended due to blood in the cochlear duct

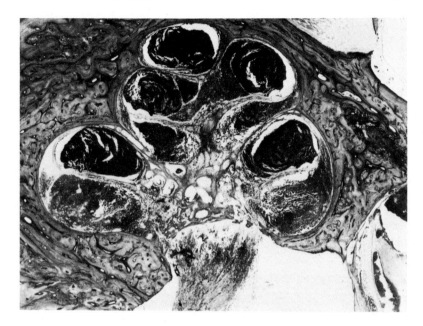

Fig. 1. Horizontal section of right ear (case 1). Blood is seen both in the endolymphatic and perilymphatic spaces. Reissner's membrane bulges. Tissue bleeding is observed in spiral ligament and modiolus.

(fig. 2). Though the tunnel and supporting elements were preserved in the organ of Corti, outer hair cells were missing. The tectorial membrane disappeared in most turns, but it was found to be lifted and deviated in other parts (fig. 3). The bleeding was seen in the spiral ligament of the basal and middle turns. The stria vascularis in the basal turn was filled with blood, and the marginal cell lining was disrupted where the blood was spouted into the cochlear duct. The modiolus and spiral ligament were filled with blood which leaked into the perilymphatic space. The saccule and perilymphatic space in the vestibule were occupied by blood. The blood had gushed from the subepithelial tissue of the saccule and the otolith were dislocated into blood mass. Only in the perilymphatic space in the pars superior of membranous labyrinth, hemorrhage was observed.

Left Ear. The blood was located mainly in the perilymphatic space of the cochlea (fig. 4). The precipitates containing a few erythrocytes were found in the apical turn. The organ of Corti, tectorial membrane and stria vascularis were normal. The blood was found only in the perilymphatic space of the vestibule with exception of the saccule which was filled with blood and bulged.

Case 2

The patient was a 46-year-old man who died of adenocarcinoma of stomach. The hematological data showed typical DIC syndrome and hemorrhagic diathesis occurred 3 months before death. 59 days prior to death, he suddenly developed a total hearing loss

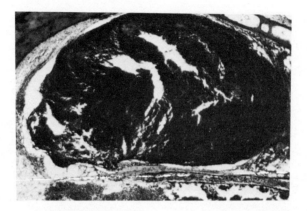

Fig.2. Basal turn of cochlea in right ear (case 1). Reissner's membrane extends due to blood in the cochlear duct.

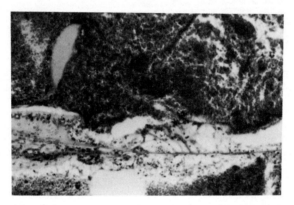

Fig.3. The tectorial membrane of right ear in case 1 is lifted by blood. Thus, it is torn away and finally lost.

in the right ear. The vestibular symptoms did not accompany it. His left hearing suddenly decreased 13 days before death. After that time the patient could not be communicated with even by a loud voice. Horizontal nystagmus to the right was observed with a Frenzel glass, though vestibular symptoms were not complained of.

Temporal Bone Findings

Right Ear. Distension of Reissner's membrane and cochlear duct were seen in the apical and middle turns (fig. 5). Reissner's membrane of the basal turn showed fluttering and tended to collapse (fig. 5). The organ of Corti was lost and flattened in all turns. Strial atrophy and severe loss of spiral ganglion cells were observed in the basal turn. The blood or its related precipitates was found only in the limited parts of the cochlear duct as is shown

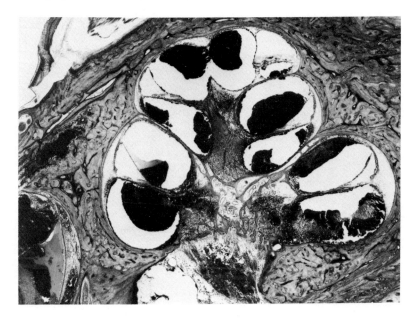

Fig. 4. Left ear of case 1. Blood is found only in perilymphatic space. Within the saccule, blood is seen.

in figure 5. The tectorial membrane was completely gone (fig. 6). The strial atrophy was seen in the basal turn (fig. 6). A thin connective tissue was seen in the scala tympani of the basal turn (fig. 5) which is possibly due to bleeding in the perilymphatic space. The saccule was filled with dissolving blood.

Left Ear. Reissner's membrane was distended in the middle and basal turns (fig. 7). Blood was found in the cochlear duct of all turns and in the scala tympani of basal turn (fig. 7, 8). Precipitates of dissolved erythrocytes were observed in the cochlear duct of apical and middle turns (fig. 7). The saccule and perilymphatic space of vestibule were occupied by a massive amount of blood. The organ of Corti was found to disappear in some turns, but it was visible in some other turns (fig. 8). The spiral ganglion cells were well preserved. The tectorial membrane was gone in most parts of the cochlea. The bleeding site could be confirmed in the stria vascularis of the basal turn.

Discussion

Inner ear hemorrhage generally occurs in the perilymphatic space [1, 2]. The site of perilymphatic bleeding was from the modiolus and/or spiral ligament facing perilymphatic space. This perilymphatic hemorrhage is

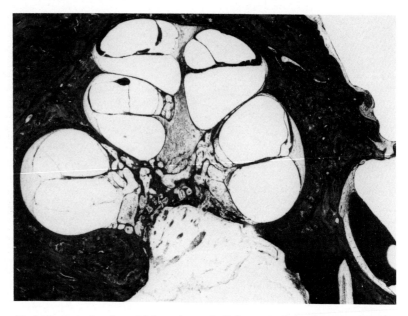

Fig. 5. Horizontal section of right ear in case 2. 59 days passed after the onset of hearing loss. The cochlear duct of apical and middle turns is expanded, but it collapses in a part of basal turn. A remnant of blood is seen in the cochlear duct. The organ of Corti is flattened. The strial atrophy is found in the most basal portion. The saccule is filled with blood.

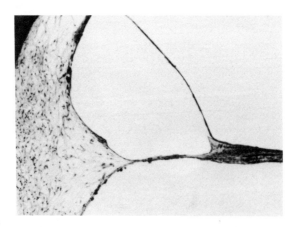

Fig. 6. Basal turn of right ear in case 2. Slightly extended Reissner's membrane, missing tectorial membrane, flattened organ of Corti and strial atrophy are seen. All these findings are thought to result from endolymphatic hemorrhage.

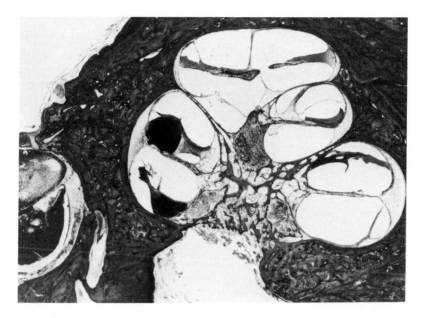

Fig. 7. Left ear of case 2. The hemorrhage occurred 13 days prior to death. The cochlear duct is expanded in the basal and middle turns. In a part of basal turn, severe endolymphatic bleeding is seen. Other parts of cochlear duct show precipitates of dissolved erythrocytes. The blood is also seen in the scala tympani, saccule and perilymphatic space in the vestibule.

known to cause severe hearing loss, speculatively due to biochemical change in the inner ear fluid [1]. However, hemorrhage into the cochlear duct is less frequently encountered. As is seen in these cases of the present study, the patients lost hearing immediately and completely which was never recovered. The site of bleeding was the stria vascularis. Firstly, the bleeding occurred in the spiral ligament which broke the strial tissues, otherwise vessels of the stria vascularis were destroyed. In both cases, the blood filled the stria which finally caused the atrophy of the stria vascularis. Blood gushing from the stria caused the tectorial membrane to be lifted and blown away. The organ of Corti was severely damaged by blood filling the cochlear duct; finally the organ of Corti became atrophied and flattened. Afterwards, this caused the secondary degeneration of cochlear nerve resulting in loss of spiral ganglion cells in Rosenthal's canal.

Reissner's membrane was distended due to blood in the cochlear duct. This blood dissolved and was thought to disappear after a certain period. The period is still under discussion. At least in case 2 the blood was hardly

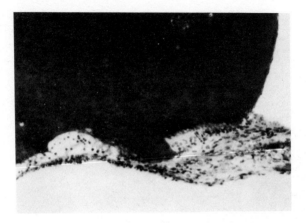

Fig.8. Case 2. Massive endolymphatic hemorrhage is observed in the basal turn. Tectorial membrane is lost, but the organ of Corti is not flattened yet.

seen in the cochlear duct of the right ear after 59 days. After the disappearance of blood, Reissner's membrane maintained its abnormal position as in endolymphatic hydrops. In case the volume of endolymph decreases, the membrane flutters or attaches to the basilar membrane and stria vascularis as endolymphatic collapse. These morphological features of Reissner's membrane vary depending on the turns of cochlea. Atrophy of the stria vascularis, which was observed in the basal turn of the right ear in case 2, resulted from the site of bleeding.

Summarizing the above findings, the endolymphatic hemorrhage shows the following pathologies after the blood disappears. They are missing tectorial membrane, flattened organ of Corti, severe loss of spiral ganglion cells, distended or fluttering Reissner's membrane and strial atrophy.

References

1   Schuknecht, H. F.: Pathology of the ear (Harvard University Press, Cambridge 1974).
2   Schuknecht, H. F.; Igarashi, M.; Chasin, W. D.: Inner ear hemorrhage in leukemia. Laryngoscope, St. Louis *75:* 662–668 (1965).

T. Ishii, MD, Department of Otolaryngology, Tokyo Women's Medical College, 10 Kawadacho, Shinjukuku, Tokyo 162 (Japan)

Adv. Oto-Rhino-Laryng., vol. 31, pp. 155–164 (Karger, Basel 1983)

Pathology as It Relates to Surgery of Menière's Disease

A. Belal, Jr.

Neurotology Section, ORL Department, Alexandria School of Medicine, Alexandria, Egypt

Introduction

Surgical procedures currently used to treat Menière's disease were developed and applied before the etiologic factors and pathogenesis of this disease were completely understood. The surgeon usually bases indications for surgery upon the patient's symptoms and failure of medical treatment to achieve symptomatic improvement. The purpose of this report is to discuss the histopathological findings in the ear following the different surgical procedures meant to achieve symptomatic improvement of Menière's symptom complex. Table I summarizes the clinical findings in the cases examined.

Endolymphatic Subarachnoid Shunt

Animal Experiments. Experimental fistulae done in different parts of the membranous labyrinth had no effect on experimental endolymphatic hydrops in guinea pigs [1]. However, *Konishi and Shea* [2] succeeded in relieving endolymphatic hydrops by experimental fistulization of the lateral semicircular canal.

Human Pathological Studies. Harrison and Naftalin [3] reported the temporal bone findings of a patient who died while undergoing endolymphatic sac procedure. The mastoid cavity was in communication with the endolymphatic sac. The inner ear showed severe endolymphatic hydrops.

Table I. Goals of surgery of Menière's disease

Case/Age/Sex	Approach	Surgery-to-death period, years	Dizziness	Hearing	Tinnitus
1/51/M	ultrasound	unknown	same	same	same
	ESS	3	none	better	same
2/66/F	ESS	5	better	same	better
3/62/F	ESS	5	same	same	worse
	revision ESS	3.5	same	same	same
	MF	2.5	none	same	better
4/50/M	cryo OPS	8	better	same	better
5/71/M	cryo OPS	13	better, 4 years	worse	unknown
6/69/F	TCL	10	constant	none	unknown
7/75/F	TCL	9	constant	none	unknown
8/61/F	TL	8	constant	none	same

ESS = Endolymphatic subarachnoid shunt; MF = middle fossa vestibular neurectomy; OPS = otic perotic shunt operation; TCL = transcanal labyrinthectomy; TL = translabyrinthine vestibular neurectomy.

Present Series. Three temporal bones belonging to 3 cases of Menière's disease who underwent endolymphatic subarachnoid shunt (ESS) procedure 3–5 years before their death are included. Cases 1 and 2 had a successful procedure with alleviation of the patient's symptoms. Case 3 had an unsuccesful procedure after initial improvement of 1 year. He later underwent a revision ESS then a middle fossa vestibular neurectomy. The details of these cases were previously reported [4]. Figure 1 summarizes the histological findings in case 3.

Ultrasonic Irradiation

Animal Experiments. Ultrasonic irradiation produces selective histological changes in the vestibular sensory areas of cats [5] and guinea pigs [6].

Human Pathological Studies. Sorrensen and Anderson [7] reported total obliteration of the lateral semicircular canal by fibrous and osteoid tissue following ultrasonic treatment. Degeneration of the superior and lateral semi-

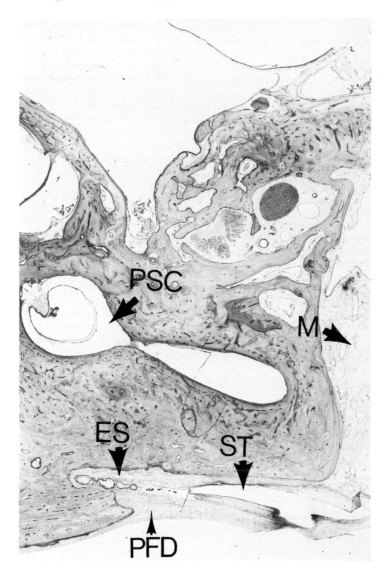

Fig. 1. Case 3, a 62-year-old female who underwent two ESS procedures that failed to relieve her symptoms. Histological examination of the temporal bone 3.5 years later showed a large mastoid operative cavity (M). The endolymphatic sac (ES) showed minimal fibrosis and was in continuity with the area occupied by the shunt tube (ST). There was no endolymphatic hydrops. PFD = posterior fossa dura; PSC = posterior semicircular canal.

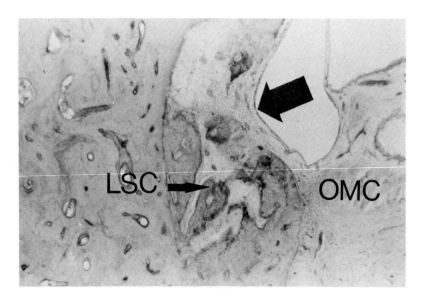

Fig. 2. Case 1, a 51-year-old male who underwent ultrasonic irradiation that failed to improve his symptoms. The non-ampullated end of the lateral semicircular canal (LSC) showed a bony defect (arrow) that communicated with the operative mastoid cavity (OMC). The lumen of this canal was obliterated by fibrous and osteoid tissue.

circular canal cristae were reported in the temporal bone of a patient with Menière's disease who underwent ultrasonic treatment [8].

Present Series. Case 1 in this series underwent ultrasonic treatment, with unknown technique and duration, that failed to alleviate the patient's symptoms. He later underwent a successful ESS operation. Figure 2 summarizes the histological findings following ultrasonic treatment in this case.

Cryogenic Otic Perotic Shunt

Animal Experiments. Fibrosis and bony obliteration of the perilymphatic space in the labyrinth were limited to the site of cryoprobe application in squirrel monkeys and pigeons [9].

Human Pathological Studies. None have been reported.

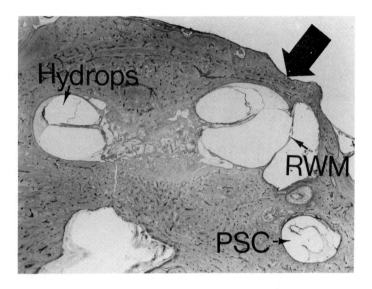

Fig. 3. Case 4, a 71-year-old male who had cryosurgery of the round window 13 years before his death. His symptoms were completely relieved postoperatively. The promontory showed a bony defect at the site of the probe application (arrow). The posterior semicircular canal (PSC) showed perilymphatic fibrosis. RWM = round window membrane.

Present Series. 2 cases of Menière's disease underwent cryosurgery in the present series. These cases were previously reported [10]. Figure 3 summarizes the histological findings in case 4.

Middle Fossa Vestibular Neurectomy

Animal Experiments. Severe degenerative changes in the vestibular labyrinth associated with endolymphatic biochemical changes were reported [11] after sectioning the superior vestibular and singular nerves and their accompanying blood vessels in cats. Similar changes were noticed in squirrel monkeys [12].

Human Pathological Studies. Most of these studies are related to surgical specimens of the vestibular nerves examined with light and electron microscopy [13].

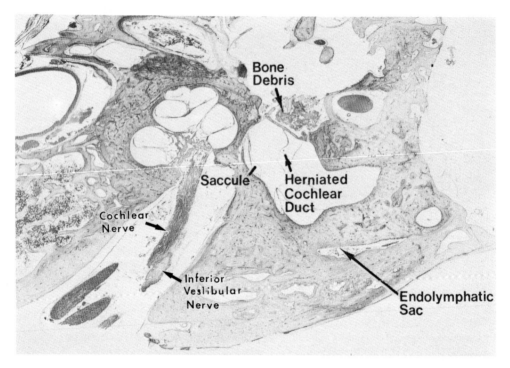

Fig.4. Case 3 who underwent middle fossa vestibular neurectomy after two failed ESS operations. There was total denervation of the vestibular end organs. The superior and inferior vestibular nerves had completely degenerated. The cochlea, cochlear and facial nerves were intact.

Present Series. 1 case underwent middle fossa vestibular neurectomy in this series. This case was previously reported [14]. Figure 4 summarizes the histopathological findings following surgery in case 3.

Transcanal Labyrinthectomy

Animal Experiments. Unilateral labyrinthectomy in cats did not cause any changes in the ipsilateral cochlear nuclei as far as the appearance of cells, volume of nuclei and the number of neurons [15].

Human Pathological Studies. Silverstein [16] reported the temporal bone findings of a case of Menière's disease that underwent transcanal labyrinthectomy 5 months before death. The vestibule was filled with fibrous tissue and

(3) Ultrasonic irradiation and cryosurgery of the labyrinth results in limited degenerative changes close to the site of probe application. Degenerated intact membranous walls may act as an internal otic-perotic shunt and relieve the symptoms of Menière's disease. The idea of *selective vestibular labyrinthectomy* and *internal shunting procedures* should be developed further.

(4) Success of the shunting procedures cannot be histologically judged by the position of Reissner's membrane. This membrane acts like varicose veins: once dilated always dilated.

(5) Recurrence of symptoms following shunting procedures may be due to failure of the shunt, or to the presence of hydrops in the contralateral ear.

(6) MF vestibular neurectomy resulted in complete denervation of the vestibular endorgans with no effect on the cochlea or facial nerve. Excision of Scarpa's ganglion resulted in retrograde degeneration of the proximal stump of the vestibular nerve.

(7) Recurrence of dizziness following transcanal labyrinthectomy occurred most commonly due to *inadequate removal of vestibular endorgans.*

(8) The high regenerative capacity of the vestibular nerve was evidenced by *traumatic neuroma* formation in the vestibule following transcanal labyrinthectomy. Whether these neuromas produce symptoms is unknown.

(9) Persistent severe cochlear hydrops following transcanal labyrinthectomy and translabyrinthine vestibular neurectomy may be the cause for the persistent tinnitus and pressure sensation.

References

1   Kimura, R.; Schuknecht, H.F.: Effect of fistulae on endolymphatic hydrops. Ann. Otol. Rhinol. Lar. *84:* 271–286 (1975).

2   Konishi, S.; Shea, J.J.: Experimental endolymphatic hydrops and its relief by interrupting the lateral semicircular canal in guinea pigs. J.Lar.Otol. *99:* 577–592 (1975).

3   Harrison, M.S.; Naftalin, L.: Menière's disease. Mechanism and management (Thomas, Springfield 1968).

4   Belal, A.A.; Honse, W.F.: Histopathology of endolymphatic subarachnoid shunt surgery for Menière's disease. Am. J. Otol. *1:* 34–44 (1979).

5   Bajek, M.: Ultrasound for Menière's disease. Archs Otolar. *97:* 133–134 (1973).

6   Lundquist, P.G.; Schindler, R.A.; Stahle, J.: Ultrasonic irradiation of the guinea pig labyrinth. Acta oto-lar. *85:* 85–95.

7   Sorrensen, H.; Anderson, M.S.: Long-term results of ultrasonic irradiation in Menière's disease. Clin. Otolaryngol. *4:* 125–129 (1979).

8   Bertrand, R.; Peron, D.: Histopathological observations of ultrasound treatment in Menière's disease. Case report.

9   Wolfson, R.J.; Ischiyama, E.: Current status of labyrinthine cryosurgery. Laryngo-
     scope, St.Louis *84:* 757–765 (1974).

10   Belal, A.A.; Honse, W.F.; Antunez, J.C.: Histopathology of cryosurgery for Me-
     nière's disease. Am. J. Otol. *1:* 147–150 (1980).

11   Silverstein, H.; Makimoto, K.: Superior vestibular and singular nerve section. Animal
     and clinical studies. Laryngoscope, St.Louis *83:* 1414–1432 (1973).

12   Filippone, M.V.; Igarashi, M.; Miyata, H.; Coats, A.C.; Alford, B.: Superior vesti-
     bular nerve sectioning. Experimental studies in squirrel monkeys. Archs. Otolar. *10:*
     241–245 (1975).

13   Ylikoshi, J.: Morphologic features of the normal and 'pathologic' vestibular nerve
     of man. Am. J. Otol. *3:* 270–273 (1982).

14   Belal, A.A.; Linthimm, F.H.; Honse, W.F.: Middle fossa vestibular nerve section.
     A histopathological report. Am. J. Otol. *1:* 72–79 (1979).

15   Hall, J.G.: Pathological changes in second-order auditory neurons after noise expo-
     sure and after peripheral denervation. Scand. Audiol. *4:* 31–37 (1975).

16   Silverstein, H.: Transmeatal labyrinthectomy with and without cochleovestibular
     neurectomy. Laryngoscope, St.Louis *86:* 1777–1791 (1976).

17   Golding-Wood, P.: The role of sympathectomy in the treatment of Menière's disease.
     J. Lar. Otol. *83:* 741–770 (1969).

18   Ylikoshi, J.; Belal, A.A.; Honse, W.F.: Morphology of the cochlear nerve after
     labyrinthectomy. Acta oto-lar. *91:* 161–171 (1981).

19   Ylikoshi, J.; Belal, A.A.: Human vestibular nerve morphology after labyrinthectomy.
     Am. J. Otol. *2:* 81–91 (1981).

20   Honse, W.F.; Belal, A.A.: Translabyrinthine surgery. Am. J. Otol. *1:* 189–198 (1980).

A. Belal, Jr., MD, Otorhinolaryngology Department, Alexandria School of Medicine,
Alexandria (Egypt)

Adv. Oto-Rhino-Laryng., vol. 31, pp. 165–174 (Karger, Basel 1983)

Fenestration Operation: Long-Term Histopathological Findings

Mario Fabiani

I Cattedra di Clinica Otorinolaringoiatrica dell'Università degli Studi di Roma, Roma, Italia

Introduction

The fenestration operation was performed to restore cochlear stimulation by making an opening in the labyrinth in the lateral semicircular canal thus replacing the function of the oval window obliterated by otosclerosis. To fulfil this condition, the new opening had to be placed in the wall of the scala vestibuli so that the fluid movements generated from the sound pressure on this new fenestra could involve the basilar membrane in their progress to the round window. Another necessary condition was to create a phase opposition between the two labyrinthine openings and this was realized by maintaining the round window in a cavum tympanicum and the neo-fenestra unprotected in the mastoidectomy cavity. Further phase opposition was created increasing the inertial mass of the tympanic membrane by leaving it attached to the malleus.

This surgical procedure theorized by *Bárány* [2], pioneered by *Holmgren* [6] and *Sourdille* [15] in the 1920s and 1930s, attained its classical one-stage form in the hands of *Lempert* [7,8], and in the 1940s and 1950s was the unique, widely performed treatment for hearing loss caused by otosclerosis.

The fenestrated ears could really never attain the closure of the air-bone gap, due to lack of the amplification exerted by the ossicular chain. Long-term audiometric follow-ups have shown that the patients who underwent fenestration and obtained practical hearing varies from 34% [5] to 74% [1] with a mean long-term postoperative air-bone gap of about 26 dB [10].

Stapes surgery replaced the fenestration operation in the 1960s due to its better functional results, easier technique and fewer postoperative complications. Complications and failures [4] peculiar to the fenestration operation were: occurrence of tympanic perforation when sectioning the malleus, post-operative acute serous labyrinthitis and closure of the neo-fenestra. To these had to be added those complications due to creation of a mastoidectomy cavity as skin graft necrosis, VIIth nerve damage and chronic inflammatory reaction of the mastoid cells [17].

This paper presents long-term histopathological findings in the temporal bones of subjects who underwent the fenestration operation for relief of otosclerotic hearing loss. Previous reports in the literature were found referring to post-fenestration short and medium-term findings [7, 18].

Material

Six temporal bones with fenestration from 5 subjects were found in the collection of 1,200 temporal bones of the Massachussets Eye and Ear Infirmary. One case with unilateral fenestration (patient 1) was previously reported [19]. In 1 of the 5 patients the operation was performed on both ears (patient 4). The temporal bones were removed from 6 to 18 h post-mortem with a bone-plug saw. Fixation, decalcification, celloidin embedding, serial horizontal sectioning, staining and mounting were performed as the standard method described by *Schuknecht* [13]. The 5 patients in this study had the fenestration during the period 1945–1954, at an age ranging from 18 to 45 years (average 37 years) by otologists in New York and Boston. One of the patients (patient 4) was operated bilaterally. All the labyrinths were fenestrated in the lateral semicircular canal overlaying the ampulla (fenestra nov-ovalis by Lempert's technique). The postoperative survival time ranged from 11 to 32 years. The causes of death were malignancies in the four females and massive intracranial hemorrhage in the male. 3 of the 5 subjects had stapedectomy in the contralateral ear (patients 1, 3 and 5).

Case Histories

The clinical histories and the histopathological reports are presented following the order by increasing length of survival time after surgery.

Patient 1, a housewife, died at age 51 (11 years after the fenestration operation).

Otologic History. At the age of 37 the patient complained of having had a slowly progressive bilateral hearing loss for at least 11 years which was diagnosed as otosclerosis. 3 years later, after a failed right stapes mobilization, a fenestration operation was performed on that ear and the auditory threshold improved.

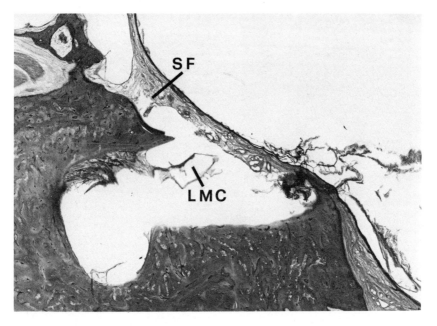

Fig.1. Patient 1. Right ear. Fenestra nov-ovalis on the ampullary end of the lateral semicircular canal. Skin flap (SF) lifted off the anterior margin of the fenestra. In this particular section the lateral membranous canal (LMC) appears torn and collapsed due to artifactual changes as could be verified in adjacent sections.

Histopathology (Right Ear). Otosclerotic foci involve the anterior margin of the oval window together with the corresponding margin of the stapedial footplate and the margins of the round window partly invading the adjacent scala tympani. The crura of the stapes are fractured and the posterior half of the tympanic membrane is retracted and adherent to the head of stapes. The walls of the mastoidectomy cavity are lined with a dense layer of connective tissue covered by squamous epithelium and the residual mastoid cells are lined with a thick layer of loose fibrous tissue with central cystic spaces containing fluid. There is a large fenestra nov-ovalis (fig. 1) overlayed by a skin flap partly pulled away from the anterior margin and it cannot be determined with certainty whether this is an antemortem condition or postmortem artefact. The underlying lateral semicircular membranous canal and ampulla appear intact in serial sections. Sense organs and auditory and vestibular nerves appear normal.

Patient 2, a housewife, died at age 56 (18 years after the fenestration operation).

Otologic History. At the age of 29 a diagnosis was made of otosclerosis and 9 years later a fenestration operation was performed on the right ear following which she experienced a permanent improvement in hearing. She visited an otologist on numerous occasions for removal of crust from her fenestrated cavity.

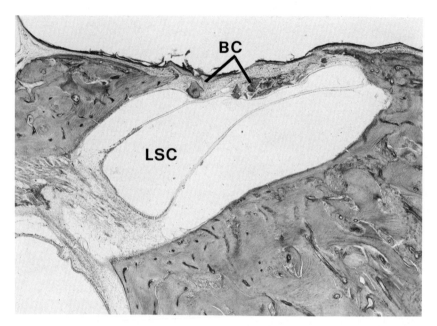

Fig.2. Patient 2. Right ear. Fenestra nov-ovalis on the ampullary end of the lateral semicircular canal (LSC). There are devitalized bony dust particles (BC = bony chips) in the thin connective tissue layer that bridges the surgically created defect on the bony wall.

Histopathology (Right Ear). There is a large focus of otosclerotic bone at the anterior margin of the oval window involving the entire footplate of the stapes and ankylosing it in the window. A small otosclerotic focus can be found in the anterior margin of the internal auditory canal. The tympanic membrane is displaced medially, and thickened. The posterior wall of the mastoidectomy cavity is approximately 6 mm in thickness and consists of widely spaced bone trabeculae separated by cells containing a thick, fibrous, submucosa with small central cystic spaces lined by epithelium. There is no evidence of osteoneogenesis in the remaining mastoid. The surface lining the surgically created mastoid bowl consists of a thin layer of squamous epithelium on a very thin layer of connective tissue. There is no evidence of ulceration or inflammatory reaction. There is a fenestra nov-ovalis measuring 2 mm in its greatest diameter (long axis of the canal) with some devitalized bone chips in the connective tissue layer which bridges the bony defect (fig.2). The underlying membranous semicircular canal is intact. The organ of Corti, vestibular sense organs, and the auditory and vestibular nerves appear normal throughout.

Patient 3, a male pharmacist, died at age 63 (18 years after fenestration operation).

Otologic History. At about the age of 35, the patient first noted progressive hearing loss, subsequently diagnosed as otosclerosis. At the age of 45 a fenestration operation was performed on the left ear. The hearing was improved for about 6 years but gradually deteriorated. Following fenestration there was recurring discharge from the left mastoid cavity.

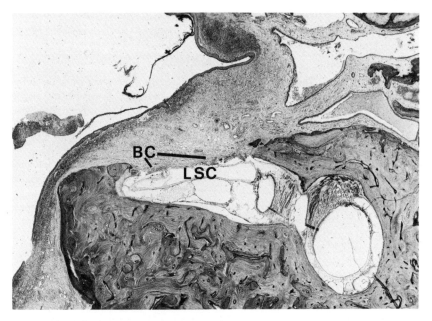

Fig.3. Patient 3. Left ear. Fenestra nov-ovalis on the ampullary end of the lateral semicircular canal (LSC). The window is covered by a 2-mm thick membrane consisting of dense highly cellular fibrous tissue surfaced with squamous epithelium. Bony chips (BC) are present in the perilymphatic space of the canal and on the under surface of the soft tissue in the window.

At the age of 60, examination revealed the left fenestration cavity to contain moist epithelial debris. The right tympanic membrane appeared normal. Audiometric studies showed combined sensorineural and conductive hearing loss.

Histopathology (Left Ear). There is a moderately large fenestration cavity partially lined by squamous epithelium. In some areas there is a thick layer of highly cellular fibrous tissue with a non-epithelialized fibrous surface and in one area there is a small granuloma. Underlying this surface tissue are many cystic spaces filled with a basophilic fluid. There is resorption of some of the underlying bone characteristic of osteitis. The posterior half of the footplate has been fractured and displaced 3 mm into the vestibule. The anterior half of the footplate is in normal position and ankylosed to the wall of the oval window by an otosclerotic focus.

There is a fenestra nov-ovalis (fig.3) in the horizontal semicircular canal overlying the ampulla and part of the membranous canal. This window measures approximately 3 mm in diameter and is covered by a very thick membrane (about 2 mm) consisting of dense, highly cellular fibrous tissue covered by squamous epithelium. There is bone dust in the perilymphatic space of the canal and on the under surface of the soft tissue in the window. There is no evidence of osteogenesis. The underlying membranous labyrinth is intact. There is some loss of spiral ganglion cells in the basal turn of the cochlea. However, the organ of Corti and its hair cell population appear normal. The stria vascularis appears normal.

Patient 4, a housewife, died at age 64 (22 years after the fenestration operations).

Otologic History. This patient first noted bilateral hearing loss at the age of 12. At the age of 42 she had fenestration operations followed by revisions in both ears by Dr. *Lempert* in New York. Audiometric studies carried out after 20 years showed a severe combined sensorineural and conductive hearing loss in both ears, much worse on the right. In the left ear the air-bone gap appeared to be about 25 dB and speech discrimination of 60%.

Histopathology. The otosclerotic lesions and the postoperative findings appear almost identical and the ears can be described together with certain small differences being pointed out.

There are large otosclerotic foci involving the cochlear walls and the oval window and round window areas. The footplate and anterior crus of both stapes are greatly thickened and firmly fixed at the margins of the oval window by otosclerotic bone. There is a separate otosclerotic focus surrounding part of the crus commune. The inferior part of the tympanic membrane of the right ear appears to have been traumatized, evidenced by the accordion-like folding of the pars propria in the anterior region and the presence of a thin replacement membrane posteriorly. The mastoid bowls are lined with a layer of fibrous tissue which is surfaced by a thin layer of squamous epithelium.

The peripheral mastoid air cells are obliterated with fibrous tissue and central cystic spaces. The lateral semicircular canals have been predominantly exposed to the mastoid cavities by the removal of the surrounding bone and there are surgical fenestrae nov-ovales (fig. 4a, b).

The windows are bridged by thick layers of fibrous tissue (about 1 mm) and surfaced by squamous epithelium. There is bone dust in the ampullae, some of which is embedded in fibrous tissue. The ampullary walls in the region of the fenestrae are torn and collapsed. Posterior to the ampullae, the canals are filled by fibrous tissue and bone with obliteration of the membranous canals for a distance of 3–4 mm. Further posteriorly, the endolymphatic canals are present and surrounded by bone and fibrous tissue extending almost to the non-ampullated ends. The cristae of the superior and lateral semicircular canals appear to have decreased hair cell populations. The saccular and utricular maculae and posterior canal cristae appear normal. There is patchy atrophy of the stria vascularis in all three turns of both ears. No endolymphatic hydrops can be seen in either ear. There is a loss of about 90% of the spiral ganglion cells in turn 1, 50% in turn 2 and 10–20% in turn 3.

Patient 5, a female whose occupation was unknown, died at the age of 49 (32 years after the fenestration operation).

Otologic History. At the age of 17 the patient noted a bilateral hearing loss and tinnitus. In the same year a fenestration operation on the lateral semicircular canal was performed on the right ear. The fenestra was measured to be 4.5 × 1.5 mm in size. The operative report states that a second small fenestra of 1.5 mm was made unintentionally posterior to the first fenestra in polishing the perilabyrinthine cells. Postoperatively she had nausea, vomiting and left beating nystagmus that subsided in 24 h. Hearing was improved but became worse 5 months later.

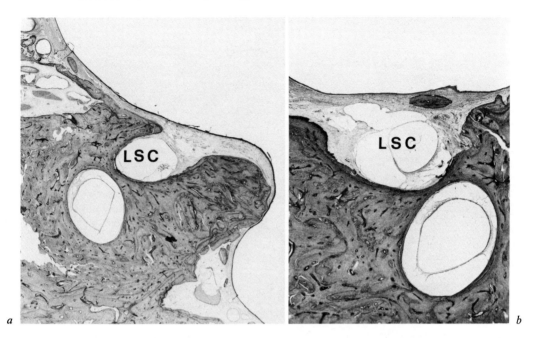

Fig. 4. Patient 4. *a* Right ear, *b* left ear. Fenestra nov-ovales on the lateral semicircular canals (LSC). Both the windows are covered by a 1-mm thick fibrous membrane.

At the age of 19, 16 months following the first operation, she underwent a revision operation on the right ear. The fenestrated areas were found to be filled with new bone which had grown into the lumen of the perilymphatic space. Some of this new bone was removed; however, the membranous labyrinth was not visualized. The second operation did not improve her hearing.

Histopathology (Right Ear). The fenestration cavity is lined with healthy appearing squamous epithelium on a thin layer of connective tissue. There is a large otosclerotic focus in the oval window involving the entire footplate and the adjacent bony cochlear labyrinth anterior to it. The niche of the round window is almost totally obliterated by a second otosclerotic focus and the residual opening is closed by thick fibrotic tissue. A third focus showing partial excavation is seen in the anterior wall of the internal auditory canal which extends circumferentially to partly encircle the cochlea but does not involve the endosteum. There is a surgical fenestra of the canal and ampulla of the lateral semicircular canal. The fenestra is obliterated by connective tissue and bone (fig. 5) for an average width of about 1 mm. The adjacent lateral canal crista shows some atrophic change. The external arm of the horizontal semicircular canal is completely filled with bone and fibrous tissue in an area posterior-inferior to the fenestra nov-ovalis, probably corresponding to the accidental opening realized during the first operation, and accounts for the postoperative peripheral vestibular syndrome.

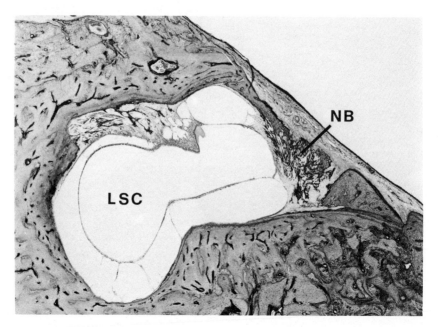

Fig. 5. Patient 5. Right ear. Fenestra nov-ovalis on the ampullary end of the lateral semicircular canal (LSC). The fenestra is obliterated by connective tissue and new bone (NB) for an average width of about 1 mm.

The organ of Corti is intact and hair cell population appears to be normal except for a loss of about 30% of cochlear neurons at the basal end of the cochlea. The utricular and saccular maculae and cristae appear to be normal. Lateral ampullary branch and utricular nerve show atrophy.

Discussion

According to *Lempert* [8] the aim of the fenestration operation was to obtain rehabilitation, for social and economic contacts, of hearing function. In the cases here examined the results were definitively positive in patients 1 and 2. Patient 3 had a 6-year hearing improvement period, followed by a gradual deterioration. This functional finding seems to be explained by the presence of a thick fibrous layer covering the neo-fenestra. Osseous closure of the fenestra is the reason for failure in restoring the hearing function in patient 5. About this patient's operation the surgeon wrote in his report: 'Fenestra made with a large polishing burr in the old technique. No lid used.'

Maintaining a long-term patency of the fenestra was one of the principal problems of the fenestration operation. *Holmgren* [6] and *Lempert* [8] continuously tried new methods to avoid ossification of the opening before attaining satisfactory technical results. It has to be underlined that patient 5 was operated in 1945 (first chronologically in the present series) when the intervention was still being perfected.

The bilateral profound hearing loss found in patient 4 can be correlated with the presence of extraordinarily aggressive pericochlear otosclerotic foci. This is an unusual finding especially in the light of the review of 103 fenestration cases [11].

One of the reasons that the fenestration procedure was replaced by stapes surgery, aside from better functional results, was the patients' need for constant care for the fenestration cavity. Some degree of mastoid inflammatory involvement is evident in the fenestrated ears, as previously described by *Pettigrew* [12]. In patient 3 osteitis and a granulomatous reaction involving the hypotympanum are present. These findings are consistent with the clinical history of recurring discharge from the cavity.

To conclude, it can be stated that the correct creation of a fenestra novovalis in expert hands was a procedure that could permanently restore some amount of hearing, without creating significant long-term damage to the inner ear cellular and neural structures.

The revival of the stapes surgery by *Shea* [14] who significantly named his technique 'fenestration of the oval window' and technical revision by *Kos, Schuknecht* and *House* caused a marked decrease in the number of fenestration operations. Still, in the middle 1960s it was pointed out [9, 16]: 'There is a place for fenestration of the horizontal canal in the treatment of conductive deafness', in particular for otosclerosis when an obliterative footplate is present, in congenital stapes anomalies and in chronic otitis media after tympanoplasty type III and type IV when there is a moveable drumhead and an intact cavum minor. In more recent years it was proposed [3] a two-stage fenestration tympanoplasty as a 'form of total tympanoplasty type V that incorporates the mastoido-tympanomastoidectomy (radical mastoidectomy), and revives the Lempert horizontal semicircular canal fenestration operation'.

Acknowledgements

The author is deeply indebted to Prof. *H. F. Schuknecht*, Chairman of Otolaryngology in the Massachusetts Eye and Ear Infirmary for his advice in preparing this paper and for providing photo-documentation.

This study was carried out during a 6-month period spent in research in Massachusetts Eye and Ear Infirmary, Department of Otolaryngology, Harvard Medical School, USA.

The *Annals of Otology, Rhinology and Laryngology* kindly permitted the reproduction of figure 1.

References

1 Adin, L.E.; Shambaugh, G.E., Jr.: A study of long-term hearing results in fenestration surgery. Archs Otolar. *53:* 243–249 (1951).

2 Bárány, R.: Die Indikationen zur Labyrinthoperation. Acta oto-lar. *6:* 260–283 (1924).

3 Blatt, I.M.: Fenestration tympanoplasty: an adjunctive technique for hearing restoration. Otolaryngol. Head Neck Surg. *87:* 366–371 (1979).

4 Danic, J.S.: Chirurgie de la surdité: l'opération de Lempert, pp. 83–89. (L'expansion scientifique française, Paris 1948).

5 House, H.P.: Long-term results in fenestration surgery. Ann. Otol. Rhinol. Lar. *60:* 1153–1159 (1951).

6 Holmgren, G.: The surgery of otosclerosis. Ann. Otol. Rhinol. Lar. *46:* 3–21 (1937).

7 Lempert, J.: Fenestra nov-ovalis: a new oval window for the improvement of hearing in cases of otosclerosis. Archs Otolar. *34:* 880–893 (1941).

8 Lempert, J.: Lempert fenestra nov-ovalis with mobile stopple. Archs Otolar. *41:* 1–41 (1945).

9 Martin, H.; Gignoux, M.; Cajgfinger, H.: La fenestration garde-t-elle encore des indications? J. Fr. Otorhinolaryngol. *15:* 527–530 (1966).

10 Masciotra, N.J.; Caparosa, R.J.: A comparison of fenestration of the horizontal canal and stapedectomy in the opposite ear. Laryngoscope, St. Louis *88:* 1725–1731 (1978).

11 Miyamoto, R.T.; House, H.P.: Cochlear reserve in otosclerosis. Archs Otolar. *104:* 464–466 (1978).

12 Pettigrew, A.M.: Histopathology of the temporal bone after open mastoid surgery. Clin. Otolaryngol. *5:* 227–234 (1980).

13 Schuknecht, H.F.: Temporal bone removal at autopsy. Preparation and uses. Archs Otolar. *87:* 129–134 (1968).

14 Shea, J.J., Jr.: Fenestration of the oval window. Ann. Otol. Rhinol. Lar. *67:* 365–379 (1958).

15 Sourdille, M.: Résultats primitifs et secondaires de quatorze cas de surdité par otospongiose opérés. Revue Lar. *51:* 595–612 (1930).

16 Walsh, T.E.: Fenestration in stapedectomy era. Archs Otolar. *82:* 346–354 (1965).

17 Williams, H.L.: The technique, end results and present status of the fenestration operation; in Schuknecht, Otosclerosis, pp. 221–238 (Little, Brown, Boston 1962).

18 Wolff, D.: Histopathology of fenestration, animal and human studies; in Schuknecht, Otosclerosis, pp. 259–268 (Little, Brown, Boston 1962).

19 Wolff, D.; Schuknecht, H.F.; Bellucci, R.: Otosclerosis and multiple surgery in the temporal bone. Ann. Otol. Rhinol. Lar. *77:* 37–42 (1968).

M. Fabiani, MD, Reparto di Fisiopatologia Uditiva, Clinica ORL,
Policlinico Umberto I, I–00161 Roma (Italy)

Adv. Oto-Rhino-Laryng., vol. 31, pp. 175–183 (Karger, Basel 1983)

Late Posttraumatic Rhinogenic Meningitis

Temporal Bone Findings of Meningogenic ,Labyrinthitis

Haruo Saito

Shiga University of Medical Science, Otsu, Japan

It is well-recognized that injuries of the head and face represent an important problem when these injuries are complicatied by meningitis. Fractures through the cribriform plate are particularly hazardous, since the dura cannot be separated from the bone and in this way a fistulous passage is formed, allowing the entrance of microorganisms from the nasal passages. The time that elapses between the injury and the onset of meningitis is in the majority of cases within 2 weeks of the trauma [1], but some cases have been reported where the meningitis took place long afterwards [2, 5, 8, 14].

Bacterial labyrinthitis may occur by extension from the subarachnoid space through the cochlear aqueduct and other anatomical structures that connect them with the inner ear.

We have had an opportunity to examine a pair of temporal bones which showed very early labyrinthine extension from the subarachnoid space and further extension into the middle ear exclusively restricted to the round window niche.

It is the purpose of this report to present a study of a case of late posttraumatic rhinogenic meningitis, and of the routes of extention into the labyrinth and middle ear from the subarachnoid space.

Case History

A man aged 19 appeared in the emergency ward in a hospital with fever, profuse slightly purulent nasal discharge, and headache that had started some days before. He was prescribed acetylamid, and ampicillin. 1 h afterwards he reappeared with his friend because his headache became severer. He was restless and feverish, sweating profusely. He was immediately admitted to the hospital for further examination. On examination, he

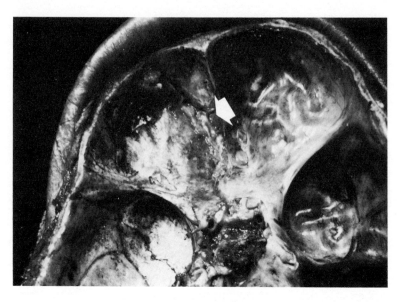

Fig. 1. A linear skull base fracture involving the left anterior cranial fossa, and evidence of massive pus accumulation along the fracture line (arrow).

fell into a stupor, but reacted to painful stimuli. No valuable information was obtained from his friend. Purulent meningitis was diagnosed and in spite of intensive therapy his condition deteriorated rapidly. On further inquiry into his past medical history, the patient's mother revealed that he had suffered a skull base fracture and fractures in the legs in a traffic accident 2 months prior to the admission, and that he was recently discharged from a hospital where he had been treated for the leg fractures. He died of the meningitis on that day.

Autopsy revealed a linear skull base fracture involving the left anterior cranial fossa, and evidence of massive pus accummulation along it (fig. 1). The temporal bones and a plug of bone, including the cribriform plate, ethomoid cells, middle and inferior turbinates, were removed for histological study, 12 h after death.

Histological Findings

Nasal Plug. There was a bony dehiscence in the cribriform plate confirming the fracture line. Subepitheliar connective tissue around this area was extremely edematous and massively infiltrated with polynuclear leuko-

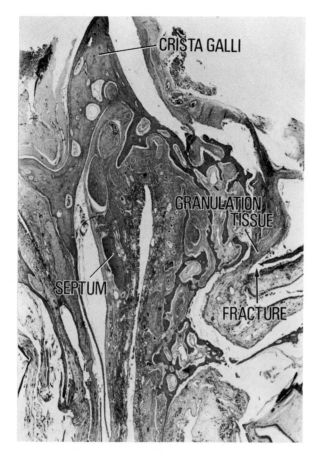

Fig.2. Small amount of granulation tissue passes through the fracture line. There are numerous leukocytes along the granulation. × 10.

cytes. There was a small amount of granulation tissue passing through the fracture line. There were numerous leukocytes along these structures (fig. 2).

Temporal Bones. The histological changes of both temporal bones resembled each other so closely that one description applies to both. There were many polymorphonuclear leukocytes in the subarachnoid space surrounding the nerve trunk in the internal auditory canal (fig. 3). The perineural and perivascular space in Rosenthal's canal contained numerous inflammatory cells increasing in number towards the basal turn. There was a

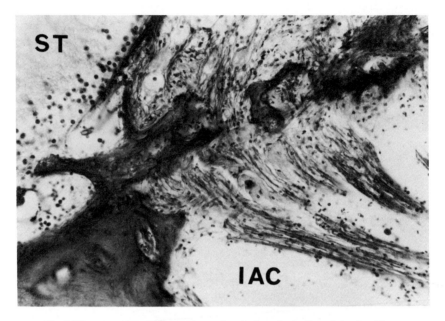

Fig. 3. There are many polymorphonuclear leukocytes in the subarachnoid space surrounding the nerve trunk in the internal auditory canal. ST = Scala tympani; IAC = internal auditory canal. × 300.

fibrinous precipitate containing polymorphonucleocytes in the scala tympani of the basal turn particularly near the orifice of the cochlear aqueduct (fig. 4). The cribrose areas in the maculae sacculi and utriculi had also polymorphonuclear leukocytes, but the amount was clearly much smaller than in Rosenthal's canal. Hair cells and supporting cells of the sense organ in the inner ear were fairly well preserved and had little degenerative change except the basal turn. The hair cells in the basal turn were both swollen and vacuolized, and protruded into the endolymphatic space, showing that some degenerative changes had taken place. The organ of Corti in the middle and apical turns remained little changed (fig. 5). Cells in some areas of the marginal and middle layers of the stria vascularis vacuolized and protruded into the endolymphatic space. There were many polymorphonuclear leukocytes on the round window membrane in the scala tympani and also a few of them on the surface of the middle ear side. The round window membrane contained inflammatory cells (fig. 6). The round window niche contained a small number of inflammatory cells. The mucosa in those areas was

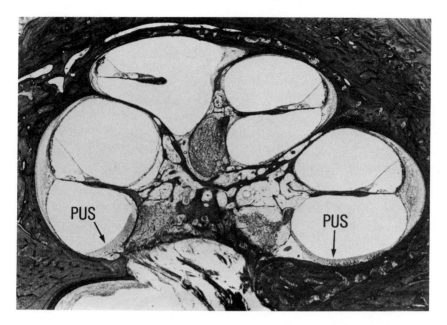

Fig.4. Perineural and perivascular space in Rosenthal's canal contains numerous inflammatory cells. × 300.

edematous and had submucous hemorrhage and capillary dilatation, revealing the acute stage of otitis media (fig. 7). The mucosa in the other areas of the middle ear and mastoid remained normal. The tympanic membranes were intact.

Discussion

The case reported here presents several interesting points, namely the interval of 2 months between the skull fracture and fulminating meningitis, the sites of involvement in this very early stage of labyrinthitis extended from the subarachnoid space, and the passage of leukocytes, in this early stage, into the middle ears through the round window membranes.

Delayed meningitis after a related head injury is rather rare. The interval between a skull fracture and the development of meningitis varies considerably. In the majority of cases it starts within 2 weeks [1], but cases have been

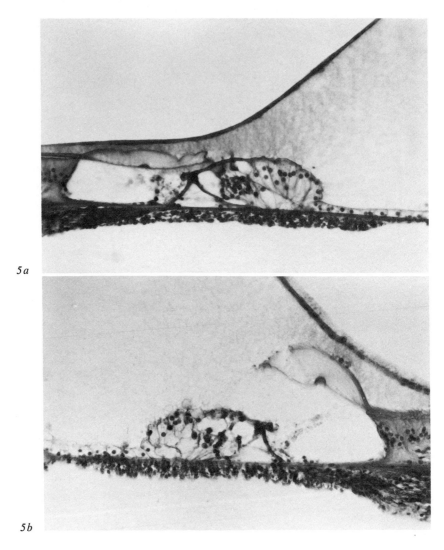

5a

5b

Fig.5. Hair cells and supporting cells in the apical and middle turns had little degenerative change *(a)*. The hair cells in the basal turn are swollen and vacuolized *(b)*. × 200

Fig.6. There are many polymorphonuclear leukocytes on the round window membrane in the scala tympani and also a few of them on the surface of the middle ear side. The round window membrane contained inflammatory cells. RWM = Round window membrane; St = scala tympani; PMN = polymorphonucleocytes. × 150.

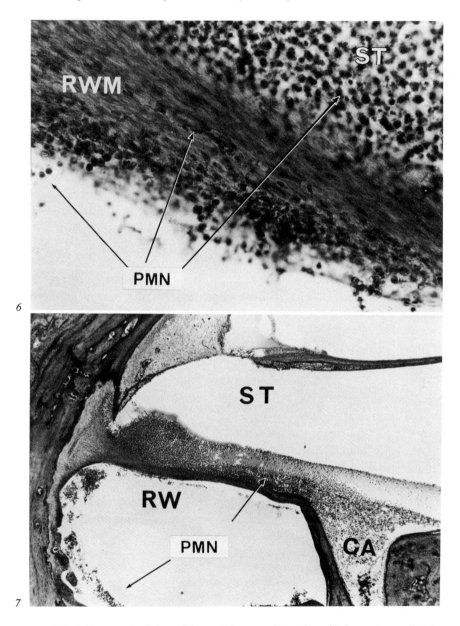

Fig. 7. The round window niche contains a small number of inflammatory cells. The mucosa is edematous and has submucous hemorrhage and capillary dilatation, revealing acute stage of otitis media. ST = Scala tympani; RW = round window; CA = cochlear aqueduct; PMN = polymorphonucleocytes. × 12.

reported in which the interval was as long as several years [1,2,5,8,14]. Recurrent attacks of meningitis have also been reported [1,2,5,14]. 1 case reported by *Grubbauer and Schmidberger* [5] had as many as 5 episodes of recurrent meningitis in a period of 5 years after a traumatic fracture of the ethomoid. Histopathology of this case report showed that the fracture through the lamina cribrosa served as a permanent path between the nasal cavity and the subarachnoid space. In addition, on reading the patient's record it is reasonable to assume that increase of intranasal pressure caused by blowing his nose made the microorganisms enter into the intracranial space by this wide opening.

Sensorineural hearing loss associated with bacterial meningitis has been described by numerous authors [3,7,9,11,13]. The incidence of hearing loss varies from 6% [9] to 21% [11], and the loss can be partial or total. *Ravio and Koskiniemi* [13] stated that the choice of initial antibiotics played a role in the sequelae of the hearing loss. Histopathological findings in cases of bacterial meningitis have been previously reported [6,7,12]; however, presence of middle ear infection in most of these cases made it difficult to identify the exact routes of extension to the labyrinth. Findings of this report showed that the infection extended exclusively from the cranial side. Pus cells were found not only in the cochlear aqueduct including the opening into the scala tympani, but also in the perineural and perivascular space connecting with the basal turn. Proportional to the invasion, the hair cell degeneration was more marked in the basal turn. For this reason we can assume that the sensorineural hearing loss as a sequelae of bacterial meningitis in its very early stage would be of high tone tilt.

Another point of special interest is the presence of pus cells in the round window membrane and in the niche. As there was no inflammatory reaction in other areas of the middle ear, localized acute otitis media in the round window niche revealed that pus cells passed through the round window membrane from the inner ear, just as the round window membrane serves as a potential route of inflammatory extension from the middle ear to the labyrinth [4,10].

References

1   Appelbaum, E.: Meningitis following trauma to the head and face. J. Am. med. Ass. *173:* 1818–1822 (1960).
2   Beks, J.W.F.: Posttraumatic nasal liquorrhoea. Arch. Chirurg. Neerland. *19:* 245–253 (1962).

3   Berlow, S.J.; Caldarelli, D.D.; Matz, G.J.; Meyer, D.H.; Harsch, G.G.: Bacterial meningitis and sensorineural hearing loss: a prospective investigation. Laryngoscope, St. Louis *90:* 1445–1452 (1980).

4   Goycoolea, M.V.; Paparella, M.M.; Juhn, S.K.; Carpenter, A.-M.: Oval and round window changes in otitis media. Potential pathways between middle ear. Laryngoscope, St. Louis *99:* 1387–1396 (1980).

5   Grubbauer, H.M.; Schmidberger, H.: Rekurrierende Meningitis purulenta bei traumatischer Siebbeinläsion. Wien, klin. Wschr. *89:* 520–524 (1977).

6   Harada, T.; Sando, I.; Myers, E.N.: Temporal bone histopathology in deafness due to cryptococcal meningitis. Ann. Otol. *89:* 630–636 (1979).

7   Igarashi, M.; Schuknecht, H.F.: Pneumococcic otitis media, meningitis, and labyrinthitis. Archs, Otolar. *76:* 126–130 (1962).

8   Jamieson, K.G.; Yelland, J.D.N.: Surgical repair of the anterior fossa because of rhinorrhea, aerocele, or meningitis. J. Neurosurg. *39:* 328–331 (1973).

9   Keane, W.M.; Potsic, W.P.; Rowe, L.D.; Konkle, D.F.: Meningitis and hearing loss on children. Archs Otolar. *105:* 39–44 (1979).

10 Mayerhoff, W.L.; Shea, D.A.; Scott, G.: Experimental pneumococcal otitis media: a histopathologic study. Otolaryngol. Head Neck Surg. *88:* 606–612 (1980).

11 Nadol, J.B.; Jr.: Hearing loss as a sequela of meningitis. Laryngoscope, St. Louis *88:* 739–755 (1978).

12 Perlman, H.B.; Lindsay, J.R.: Relation of the internal ear space to the meninges. Archs Otolar. *29:* 12–23 (1939).

13 Ravio, M.; Koskiniemi, M.: Hearing disorder after *Haemophilus influenzae* meningitis. Archs Otolar. *104:* 340–344 (1978).

14 Sengupta, R.P.; Gravan, N.: Recurrent fulminating meningitis 20 years after head injury. J. Neurosurg. *41:* 758–761 (1974).

H. Saito, MD, Shiga University of Medical Science, Otsu 520–21 (Japan)

Adv. Oto-Rhino-Laryng., vol. 31, pp. 184–190 (Karger, Basel 1983)

Pathological Findings and Surgical Implications in Herpes Zoster Oticus [1]

B. Etholm, H. F. Schuknecht

Vestfold Sentral-sykehus, Tønsberg, Norway;
Department of Otolaryngology, Massachusetts Eye and Ear Infirmary,
Boston, Mass., USA

Herpes zoster oticus is characterized by severe pain in the ear or mastoid area followed soon thereafter by a vesicular eruption of the ear canal and/or auricle and facial palsy. Auditory and/or vestibular symptoms may or may not occur. The purpose of this paper is to present the histopathological findings in the temporal bones of a patient with this disorder.

Case Report

At the age of 70, audiometry showed a bilateral sensorineural hearing loss characterized by descending audiometric patterns, slightly worse in the left ear. Speech discrimination scores were 72% on the right and 40% on the left. At the age of 71 he was admitted to the hospital because of fatigue, and the diagnosis of acute myelocytic leukemia was made. He was treated with Oncovin, Methotrexate, Purinethal, and Prednisone with some improvement. Several blood transfusions were given. 2 months later he developed complete right facial palsy followed by vesicular lesions of the right auricle and soft palate. The medical records do not indicate whether he experienced vertigo or further hearing loss. Audiometry showed a bilateral sensorineural hearing loss, somewhat more severe than 18 months previously, and worse in the right ear. Additionally, there was now a 30- to 40-dB conductive loss in the right ear. Speech discrimination was not tested. He died 2 months after the onset of the facial palsy, and both temporal bones were removed for histological study.

Histopathology

Right Ear. The specimen is in an excellent state of histological preservation and preparation. The tympanic membrane and ossicles are intact. The mucous membrane of the middle ear is thickened by fibrous tissue proliferation and infiltration with inflammatory cells, and there are collections of polymorphonuclear leukocytes in the middle ear and

[1] This work was supported by NINCDS Grant 5 R01 NS05881-16.

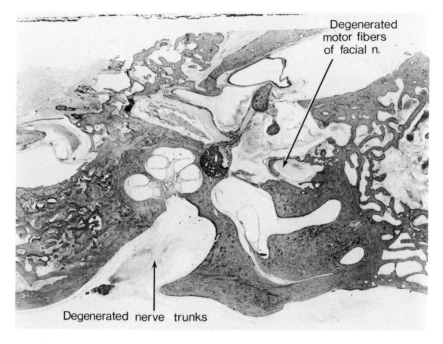

Degenerated
motor fibers
of facial n.

Degenerated nerve trunks

Fig. 1. Low power view of the right ear showing degeneration of the motor division of the facial nerve and preservation of the sensory bundle.

mastoid. There is a leukemic submucosal infiltrate in some areas. The petrous apex contains highly cellular bone marrow consistent with a myeloblastic leukemic infiltrate. There is an otosclerotic focus anterior to the oval window which is not fixing the stapes. The mastoid is well-developed and has normal trabeculation.

The internal auditory canal is filled with loose fibrous tissue, and the dura lining the canal is somewhat thickened (fig. 1). The vessels of the dura and internal auditory canal show perivascular cuffing with small round cells. In the midportion of the internal auditory canal there are small collections of extravascular erythrocytes.

The most central part of the facial nerve trunk is present and appears normal. Proceeding peripherally the nerve shows progressive degeneration which becomes complete about 5 mm from the porus acousticus (fig. 2, 3). The motor fibers of the facial nerve present three different histological features as follows: (1) centrally – normal nerve fibers, (2) midportion – atrophy with granular degeneration of nerve fibers and infiltration with small round cells, and (3) peripherally – severe degeneration of all axons, cellular debris, infiltration with round cells, and fibrosis. From the midportion of the internal auditory canal to the geniculate ganglion both the motor and sensory divisions show total degeneration. The geniculate ganglion shows a normal population of neurons and the sensory bundle distal to the geniculate ganglion is normal (fig. 4). There is total degeneration of the cochlear and vestibular nerve trunks although small numbers of Schwann cells remain.

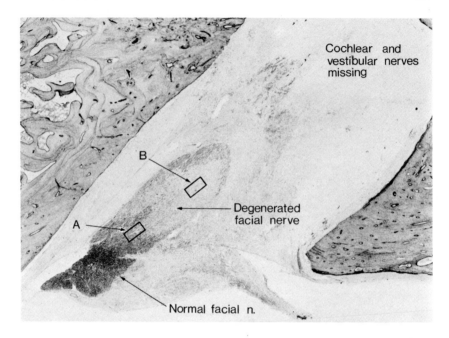

Fig. 2. Internal auditory canal of the right ear showing normal facial nerve, partially degenerated facial nerve A, and severely degenerated facial nerve B. Areas A and B are shown in figure 3.

The utricular and saccular maculae and the cristae are partially degenerated. The cupula of the lateral semicircular canal has been displaced and is surrounded by pigment-laden macrophages (fig. 5). In the most superior part of the superior semicircular canal there is a bone chip and very opaque material which cannot be clearly identified and may represent artifact of removal. This opaque material extends inferiorly into the common crus which has tears of its wall.

Left Ear. This ear is in an excellent state of histological preservation and preparation. There is removal artifact represented by tears in the tympanic membrane, dislocation of malleus and incus, and fractures of the bony external auditory canal. The petrous apex contains bone marrow which is richly infiltrated with leukemic cells. The middle ear and mastoid show no evidence of leukemic infiltrate or inflammation. The mastoid is well-developed with normal trabeculation. There is a small otosclerotic focus anterior to the oval window which is not fixing the stapes. The hair cell population of the organ of Corti appears normal except for possible slight loss at the extreme basal end of the cochlea. The stria vascularis appears normal. There is a loss of about 95% of the cochlear neurons similar to the opposite ear. The utricular and saccular maculae and the cristae appear normal. The facial nerve, geniculate ganglion and sensory bundle appear normal. The vestibular nerves appear normal.

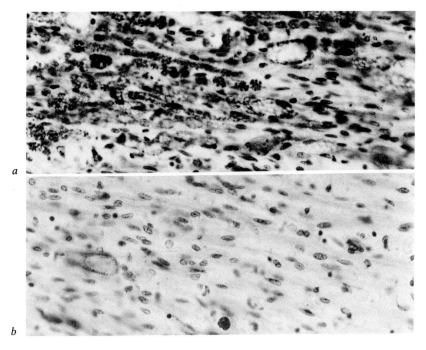

Fig. 3. Facial nerve of the right ear. *a* Area A in figure 2 showing granular degeneration of nerve fibers. *b* Area B in figure 2 showing total nerve degeneration.

Otologic Diagnoses. (1) Herpes zoster oticus with degeneration of facial, cochlear, and vestibular nerves, right. (2) Fibrous and inflammatory infiltrate, internal auditory canal, secondary to herpes zoster oticus, right. (3) Degeneration, partial, vestibular sense organs, secondary to herpes zoster oticus, right. (4) Presbycusis, neural, severe, bilateral.

Discussion

In 1907 *Ramsay Hunt* [1] suggested that herpes zoster of the face and neck with facial palsy and acoustic symptoms be considered a single disease entity. Subsequently, he [2] proposed that the underlying pathology is a posterior poliomyelitis characterized by a specific inflammation of ganglia of the spinal type. He regarded the Gasserian, geniculate, acoustic, glossopharyngeal, vagus, and second, third and fourth cervical ganglia as representing a continuous chain, all having the same embryonal origin, the neural crest.

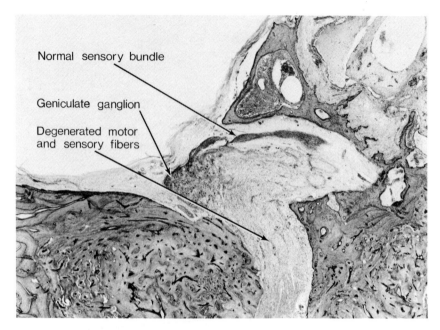

Normal sensory bundle

Geniculate ganglion

Degenerated motor
and sensory fibers

Fig.4. Normal geniculate ganglion and sensory bundle in the right ear. The motor division is totally degenerated.

Involvement of one or more of these ganglia would then cause neurological deficits and herpetic involvement in their respective areas of innervation.

It is now known that the herpes zoster virus is one of the intracellular pox viruses and that it is indistinguishable from the varicella virus which causes chicken pox. In 1965 *Hope-Simpson* [3] postulated that the virus from an original attack of varicella lies dormant in the sensory ganglia and under certain conditions (such as immunodeficiency, illness, physical stress) can be activated to cause herpes zoster lesions.

The literature contains several reports on the pathology of herpes zoster oticus as seen in autopsy material [4]. In 1934 *Maybaum and Druss* [4] found the geniculate ganglion to show slight degeneration with cytoplasmic shrinkage and pyknosis or absence of nuclei and infiltration of the ganglion with lymphocytes and plasma cells. There was also lymphocytic cuffing around blood vessels. *Sachs and House* [6] found lymphocytic infiltration in the endoneural and perineural tissues and focal collections of lymphocytes in the geniculate ganglion; however, there was no abnormality in the ganglion cells. The motor fibers of the facial nerve were severely degenerated. *Guldberg-*

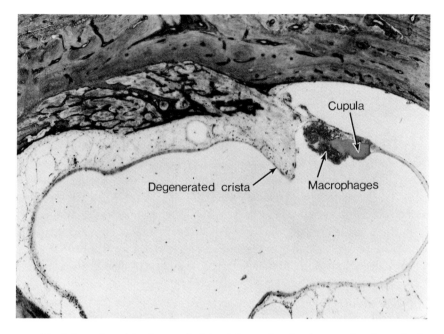

Fig. 5. Lateral semicircular canal of the right ear showing degenerative change in the crista, displacement of cupula, and macrophages with pigment.

Möller et al. [7] found the geniculate ganglion to be normal whereas the entire intratemporal segment of the facial nerve was degenerated and infiltrated with lymphocytes. In 1972, *Zajtchuk* et al. [8] presented a case of herpes zoster with vestibular symptoms and vesicular eruptions of the ear canal in a patient with Hodgkin's disease. Temporal bone study showed degeneration of the ampullary division of the superior vestibular nerve and crista of the lateral canal as well as fibrous tissue and new bone in the adjacent perilymphatic space. This same case was reported again by *Proctor* et al. [9] in 1979 with the addition of more extensive vestibular test data. *Blackley* et al. [10] found extensive lymphocytic infiltration throughout the facial, vestibular, and auditory nerves. They also noted perivascular cuffing by lymphocytes in the modiolus, perineural tissue of the facial nerve, chorda tympani and skin of the external auditory canal.

The findings in these reported cases, as well as those of our case, demonstrate the principal features of herpes zoster oticus to be: (1) variable degrees of degeneration of the facial, vestibular, and cochlear nerves, (2) preservation of the geniculate ganglion and sensory bundle of the facial nerve, (3) variable lymphocytic and plasma cell proliferation as diffuse infiltrates, focal collections and perivascular cuffs, (4) variable associated degeneration of auditory and vestibular end organs, and (5) variable extent of reparative response in the form of fibrous and osseous proliferaton.

Conclusions

(1) Herpes zoster oticus is probably caused by reactivation of a dormant varicella-like virus harbored in the auditory, vestibular, or facial nerves. (2) The inflammatory process does not originate from or involve the geniculate ganglion as originally proposed by *Ramsay Hunt* [1]. (3) When facial palsy is associated with auditory and/or vestibular symptoms, the inflammatory lesion is located in the nerve trunks within the internal auditory canal. (4) In cases with multiple cranial nerve involvement, it would seem improbable that surgical decompression of the facial nerve would be of therapeutic value.

References

1   Ramsay Hunt, J.: Herpetic inflammations of the geniculate ganglion: a new syndrome and its complications. J. nerv. ment. Dis. *34:* 73–96 (1907).
2   Ramsay Hunt, J.: A further contribution to the herpetic inflammations of the geniculate ganglion. Am. J. med. Sci. *136:* 226–241 (1908).
3   Hope-Simpson, R.E.: The nature of herpes zoster: a long-term study and a new hypothesis. Proc. R. Soc. Med. *58:* 9–20 (1965).
4   Maybaum, J.L.; Druss, J.G.: Geniculate ganglionitis (Hunt's syndrome). Clinical features and histopathology. Arch. Otolar. *19:* 574–581 (1934).
5   Findlay, J.P.: Facial paralysis: a clinical review with an autopsy report of the histopathology in a case of infection of the geniculate ganglion by the virus of cephalic herpes zoster. Med. J. Aust. *2:* 810–815 (1952).
6   Sachs, E., Jr.; House, R.K.: The Ramsay Hunt syndrome. Geniculate herpes. Neurology, Minneap. *6:* 262–288 (1956).
7   Guldberg-Möller, J.; Olsen, S.; Kettel, K.: Histopathology of the facial nerve in herpes zoster oticus. Arch. Otolar. *69:* 266–275 (1959).
8   Zajtchuk, J.T.; Matz, G.J.; Lindsay, J.R.: Temporal bone pathology in herpes oticus. Ann. Otol. Rhinol. Lar. *81:* 331–338 (1972).
9   Proctor, L.; Perlman, H.; Lindsay, J.; Matz, G.: Acute vestibular paralysis in herpes zoster oticus. Ann. Otol. Rhinol. Lar. *88:* 303–310 (1979).
10 Blackley, B.; Friedmann, I.; Wright, I.: Herpes zoster auris associated with facial nerve palsy and auditory nerve symptoms. Acta Oto-lar. *63:* 533–550 (1967).

H. F. Schuknecht, MD, Department of Otolaryngology, Massachusetts Eye and Ear Infirmary, 243 Charles Street, Boston, MA 02114 (USA)

Adv. Oto-Rhino-Laryng., vol. 31, pp. 191–197 (Karger, Basel 1983)

The Air Caloric Test and Its Normal Values

Yin-Zao Gao[a], Yu-Ying Sze[b], Lee Shen

[a] Research Fellow in Otolaryngology, Harvard Medical School
and Massachusetts Eye and Ear Infirmary, Boston, Mass., USA;
[b] Staff Members, Department of Otolaryngology, Shaanxi Provincial People's
Hospital, Xian, People's Republic of China

Introduction

Although the air caloric test for vestibular function was first introduced by *Aantaa* [1] in 1966, it was not until 1972 that *Albernaz and Gananca* [2] used it in clinical practice. Since then there has been a tendency for the air caloric test to take the place of the water caloric test [2, 3] because it has a number of advantages over water in caloric testing, including its use in patients with perforated ear drums, elimination of water collection problems [4], convenience to the operator, and better patient acceptance. *Suter* et al. [3] studied the results of water and air caloric tests in two groups of patients and concluded that the results were the same. In the experience of *Capps* et al. [5] with 175 patients, the air caloric test proved to be a most satisfactory replacement for the water caloric test in at least 95% of the cases. The remaining 5% had to be irrigated with 20 ml of ice water in order to confirm a non-functioning labyrinth.

The parameters used by otologists differ (table I) and, as far as we know, there are no normal values for the air caloric test. In 1977, we began to use the air caloric test in clinical practice. From August 1980 to May 1981, we investigated the normal values of the air caloric test. The investigation was carried out in two groups, each consisting of 100 normal persons. A detailed otological history was taken from each person to insure normal vestibular and cochlear function. Before testing, the external ear canal and ear drum were examined, and wax, if present, was meticulously removed. The age of these normal persons was between 15 and 30 years. We excluded persons above 30 years of age to avoid any influence of aging [6].

Table I. Parameters used by different authors

	Albernaz and Gananca [2]	Proctor et al. [12]	Benitez et al. [12]	Capps et al. [5]	Coats et al. [10]	Tole [8]
Temperature, ° C						
Hot air	42	50	50	50	50	45.5
Cold air	20	24	24	24	27.5	30
Time of stimulation, s	60	30	60	60	60	100
Air flow rate, l/min	not indicated	8	6	8	13	8

Method and Equipment

During testing, the subjects were seated in an examination chair with the head tilted 60 ° backward on a head rest to place the horizontal canal in the proper vertical plane (fig. 1). The apparatus used is the Air Caloric Irrigator Model LRY-1, made in China. The right and left ears are irrigated with hot and cold air alternately, altogether four times, with a 10-min waiting period between each test. We used hot stimulation first. The tip of the air caloric irrigator was directed to the posterosuperior part of the external ear canal. The depth of the irrigating tip was controlled with a mark on the irrigating tip, and was 20 mm from the outside surface of the tragus in all cases. The inside diameter of the tip was 3 mm and the outside diameter 4 mm. The persons tested were asked to fix their eyes on a target on the ceiling and nystagmus was examined with the eyes open. The latent period and duration of nystagmus were calculated with a stop watch. *Paparella* et al. [7] also routinely use the air caloric test in their clinic for vestibular evaluation and have found it to be reliable.

Results

Group I (n = 100). (1) The parameters used were: hot air at 50 °C, cold air at 20 °C, 60-second duration of stimulation, and air flow rate of 10 l/min. (2) The normal values for hot and cold air are shown in tables II and III, respectively. (3) Among the group of 100 normal persons, 3 showed directional preponderance, two to the left and one to the right.

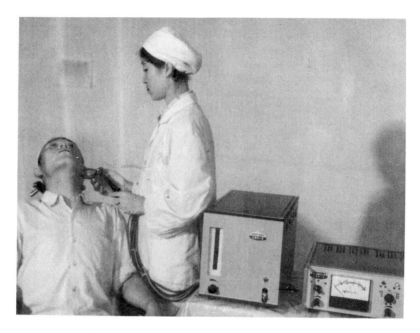

Fig.1. Position of head of person tested during examination.

Group II (n = 100). (1) The parameters used were: hot air at 46°C, cold air at 24°C, a 60-second duration of stimulation, and air flow rate of 10 l/min. (2) The normal values for hot and cold air are shown in tables IV and V. (3) In group II 2 normal persons showed directional preponderance, 1 to the left and the other to the right.

Discussion

Advantages of the Air Caloric Test. We are deeply impressed by the advantages of the air caloric test. It is easier to perform, it is tolerated better by patients, especially children, it can be used in patients with perforated ear drums or mastoid cavities, and there is no need to collect water as in the water caloric test. We now use the air caloric test as the routine vestibular function test in our clinic, and more than 500 tests have been done on normal persons and patients with various kinds of vestibular disorders.

Table II. Normal values. Hot air at 50 °C

	Left		Right		Right and left	
	latent period	duration of nystagmus	latent period	duration of nystagmus	latent period	duration of nystagmus
Mean	30.44″	1′55.9″	31.08″	1′56.4″	30.22″	1′55.2″
Standard deviation	3.43	11.98	2.93	12.23	3.22	11.91
Standard error	0.34	1.20	0.29	1.22	0.23	0.84

Table III. Cold air at 20 °C

	Left		Right		Right and left	
	latent period	duration of nystagmus	latent period	duration of nystagmus	latent period	duration of nystagmus
Mean	26.38″	2′8″	26.9″	2′11″	20.23″	2′8.4″
Standard deviation	2.69	15.50	2.28	18.05	2.79	16.37
Standard error	0.27	1.55	0.23	1.81	0.20	1.16

Factors Influencing the Results of the Air Caloric Test. Although the air caloric test has many advantages as mentioned above, it is by no means widely accepted. One major problem is its reliability. Some claim that the results of the air caloric test vary in the same subject at different times. *Tole* [8] compared the responses of the air caloric test with that of the standard water caloric test in 20 normal subjects. He came to the conclusion that any day-to-day variation in a given subject is attributable to inherent or random variability in test administration, rather than to consistent variation in the entire normal population or in the type of irrigating medium (air or water). *Paparella and Strong* [9] have tested 2,132 consecutive cases with air irrigation; excellent tolerance and good correspondence in repeated testing have been observed. We are of the opinion that the variation in the results in

Table IV. Normal values. Hot air at 46 °C

	Left		Right		Right and left	
	latent period	duration of nystagmus	latent period	duration of nystagmus	latent period	duration of nystagmus
Mean	31.58″	1′55.2″	32.56″	1′51.4″	32.07″	1′53.3″
Standard deviation	4.03	9.86	4.85	9.72	4.48	9.94
Standard error	0.40	0.99	0.49	0.97	0.31	0.71

Table V. Cold air at 24 °C

	Left		Right		Right and left	
	latent period	duration of nystagmus	latent period	duration of nystagmus	latent period	duration of nystagmus
Mean	28.22″	2′8″	28.33″	2′5.9″	28.20″	2′7″
Standard deviation	3.16	13.44	3.36	12.45	3.25	12.96
Standard error	0.13	1.34	0.34	1.25	0.23	0.92

the same subject at different times is due to technical problems, especially in the early stages when we were not quite acquainted with the maneuvers of the air caloric test.

The two main factors which influence the accuracy of the air caloric test are: (1) The depth of the tip of the air irrigator in the external ear canal must remain the same during each irrigation. *Coats* et al. [10] demonstrated that when the tip is moved outward only 7 mm, the intensity of hot air response drops by about 40% and the cold air response by about 20%. They concluded that inadvertent differences in tip position are an important source of air caloric test-retest variability. We put the tip 20 mm within the external ear canal during each irrigation. (2) The external ear canal must be dry during irrigation with air. When a wet external ear canal is irrigated with dry air, heat is taken up by evaporation of surface moisture [11]. As a result, the effect of cold air irrigation is exaggerated and the effect of hot air irrigation is diminished. Other factors influencing the results of the air caloric test are the rate of air flow, the thickness of the wall of the tip, and the presence of wax in the external ear canal.

The Presence of Directional Preponderance in Normal Persons. It is interesting to note that among the 200 normal persons, 5 (2.5%) showed directional preponderance, 3 to the left and 2 to the right. We consider directional preponderance a difference of over 40 s between the two longer durations and two shorter durations.

Parameters Recommended for Clinical Use. In selecting air irrigation parameters for clinical use, caloric response intensity, test-retest variability, and the patient's comfort should be considered. The major factor controlling the comfort of the irrigation was the flow rate. We found that air-flow rates of 10 l/min were quite well tolerated and less uncomfortable than the Hallpike water irrigation. We recommend the parameters of group II in our investigation because these parameters are enough to elicit vestibular response and are well tolerated by the subjects.

Acknowledgements

The authors are greatly indebted to Dr. *Harold F. Schuknecht* who kindly looked over and corrected the manuscript. Miss *Carol Ota* assisted in preparing and editing of the manuscript. The statistical work for this paper was done by Miss *Guang-Min Xu.*

References

1   Aantaa, E.: Caloric test with air. Acta oto-lar. suppl. *224:* 82–85 (1966).
2   Albernaz, P. L. M.; Gananca, M. M.: The use of air in vestibular caloric stimulation. Laryngoscope, St. Louis *82:* 2198–2203 (1972).
3   Suter, C. M.; Blanchard, C. L.; Cook-Manokey, B. E.: Nystagmus responses to water and air caloric stimulation in clinical population. Laryngoscope, St. Louis *87:* 1074–1078 (1977).
4   Paparella, M. M.; Strong, M. S.: The year book of otolaryngology, pp. 19–24 (Year Book Medical Publishers, Chicago 1980).
5   Capps, M. J.; Precidao, M. C.; Paparella, M. M.; Hoppe, W. E.: Evaluation of the air caloric test as a routine examination procedure. Laryngoscope, St. Louis *83:* 1013–1021 (1973).

6   Mulch, G.: Influence of age on results of vestibular function tests. Ann. Otol. Rhinol. Lar. 88: suppl. 56, pp. 1–17 (1979).

7   Paparella, M. M.; Rybak, L.; Meyerhoff, W. L.: Symposium: value and cost analysis of recent ear tests. Air caloric testing in otitis media (preliminary studies). Laryngoscope, St. Louis 89: 708–714 (1979).

8   Tole, J. R.: A protocol for the air caloric test and a comparison with a standard water caloric test. Archs Otolar. 105: 314–319 (1979).

9   Paparella, M. M.; Strong, M. S.: The year book of otolaryngology, pp. 24–26 (Year Book Medical Publishers, Chicago 1979).

10 Coats, A. C.; Herbert, F.; Atwood, G. R.: The air caloric test. Archs Otolar. 102: 343–354 (1976).

11 Proctor, L. R.: Air caloric test: irrigation technique. Laryngoscope, St. Louis 87: 1283–1390 (1977).

12 Proctor, L. R.; Metz, W. A.; Dix, R. C.: Construction of a practical and inexpensive air stimulator for caloric vestibular testing. Laryngoscope, St. Louis 86: 126–131 (1976).

13 Benitez, J. T.; Bouchard, K. R.; Choe, Y. K.: Air calorics. Ann. Otol. Rhinol. Lar. 87: 216–223 (1978).

Yin-Zao Gao, MD, Department of Otolaryngology, Shaanxi Provincial People's Hospital, Xian, People's Republic of China

Adv. Oto-Rhino-Laryng., vol. 31, pp. 198–207 (Karger, Basel 1983)

Investigation of Menière's Disorder by Extratympanic Electrocochleography

Kevin P. Gibbin, Stephen M. Mason, Biswajit Majumdar

Department of Otolaryngology and Department of Medical Physics, Queen's Medical Centre, University Hospital, Nottingham, England

Introduction

The pathological changes in Menière's disorder have been recognized since *Hallpike and Cairns* [9] first described the temporal bone findings in 1938. However, it is only with the relatively recent development of methods of detecting and recording the intracochlear and VIII nerve potentials that the electrophysiological changes have been demonstrated. *Schmidt* et al. [14] in 1974 using transtympanic electrocochleography (ECochG) found a steep slope in the input/output curves indicating recruitment and a larger summating potential (SP) relative to the action potential (AP) near threshold. *Gibson* et al. [7] have shown widening of the AP/SP waveform in Menière's disorder, the commonest ECochG finding being a broadening of the AP/SP complex due to a relative enhancement of the negative SP component; a small cochlear microphonic (CM) component is often observed.

Extratympanic ECochG has been used by a number of workers [4, 12, 15] and its use has now become established as a non-invasive technique.

Materials and Methods

We have now examined a total group of 24 patients with Menière's disorder with symptoms and neuro-otological and vestibulometric findings confirming the diagnosis. We have studied only patients with classical findings, excluding any so-called purely cochlear, purely vestibular and other atypical cases of Menière's disorder. All patients exhibited the classic triad and 16 complained of fullness in the ear at some stage usually immediately

Table I. Group mean audiogram

250 Hz	500 Hz	1 kHz	2 kHz	4 kHz
63 ± 21 dB	58 ± 20 dB	51 ± 20 dB	46 ± 17 dB	49 ± 20 dB

preceding or during an attack. The patients examined comprised 12 males and 12 females, the age range being 32–66 years, the mean age being 45.3 years.

The method used is that of extratympanic ECochG using a silver/silver chloride electrode placed in the deep meatus in a posterior-inferior position as previously described by *Mason* et al. [12]. Electrode contact impedances of 10 KΩ or less are usually achieved. An alternately inverted 100-μs click stimulus presented at 10/s is used to evoke the combined AP/SP complex from which measurements of the relative amplitude of the SP component are taken. The SP is also investigated using an alternately phased 4 kHz tone burst (8 cycles) presented at a rate of 200/s so as to achieve adaptation of the AP. The CM component is recorded using an alternately phased 1 kHz tone burst (8 cycles), the opposite phase response being inverted and summated to the inphase response so as to enhance the CM component and cancel some of the AP component.

ECochG recordings were taken both before and after administration of glycerol in the dosage of 1.5 g/kg after the method of *Klockhoff and Lindblom* [10]. Subjective pure tone audiometry was carried out at the preglycerol stage and at the end of the test.

On approximately half the patients examined (the most recently examined ones) preglycerol and postglycerol serum osmolality was measured to ensure that a dehydrating dose was given.

Results

16 of the 24 patients in the series experienced fullness in the diseased ear at some stage and in 12 of the 16, fluctuation of the hearing was recorded. The overall group mean audiogram can be seen in table I which shows an overall low tone hearing loss. The mean hearing loss as shown in table II is the average of the 4 frequencies 500 Hz, 1, 2 and 4 kHz. There appears to be no relationship between the duration of the vertigo and fluctuation of hearing

Table II. Clinical details

%SP	Threshold shift[1]	Fluctuation	Fullness	Mean hearing loss	Duration (y = years; m = months)
85	+ +	—	+	66	5 y
75	0	—	+	62	6 m
72	+ +	+	+	70	8 m
68	+ +	—	—	78	7 y
60	+	NR	+	75	3 y
59	+	—	—	54	6 m
59	0	+	+	49	9 y
57	—	+	—	54	6 y
55	0	—	—	79	2 y
53	—	+	+	34	12 y
53	+ +	NR	—	67	4 y
47	—	+	+	60	15 y
47	+	+	+	48	9 y
45	+ +	+	—	51	2 y
44	NR	+	+	66	4 y
42	0	+	+	54	1 y
41	0	+	+	48	1 y
39	0	+	+	9	3 y
37	—	+	+	62	1 y
35	—	—	+	43	4 y
34	0	+	+	31	6 m
33	0	+	—	23	20 y
32	+	+	—	60	8 y
26	0	+	+	31	8 y

[1] See text. NR = Not recorded.

– some patients with a long history experiencing fluctuation and conversely some patients with a short history of vertigo not experiencing a fluctuant hearing loss. The clinical details of patients are summarized in table II.

Figure 1 shows a typical recording and indicates how the various parameters are measured, the recording showing a typically enhanced negative SP with widening of the AP/SP complex due to this enhanced SP. Similarly the SP can be clearly seen on the 4 kHz SP run. Typical recruiting patterns of ECochG were recorded with rapid reduction of the amplitude of the N1 component with decreasing stimulus intensity.

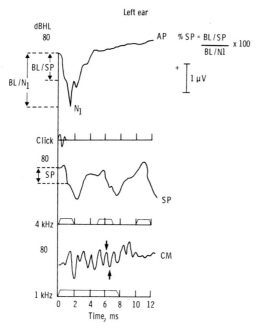

$$\% SP = \frac{BL/SP}{BL/N1} \times 100$$

Fig. 1. ECochG recording showing method of measuring CM, AP and SP.

In analyzing our results the preglycerol recordings are considered in addition to the changes induced by glycerol dehydration. The range of amplitude of both the BL/N1 and the BL/SP is large, 0.56 to 6.05 and 0.15 to 2.83 μV, respectively. As previously reported [6], it has been found more useful to express the amplitude of the SP component as a percentage of the AP/SP complex. The percentage SP in our series ranged from 26 to 85%, the distribution of percentage SP being shown compared with normals in figure 2. The mean and standard deviations of the percentage SP are 50 ± 15% for the Menière's group compared with 22 ± 12% in our normal series. There is seen to be considerable overlap in the range for normals and Menière's cases as can be seen in figure 2. The amplitude of the percentage SP is related to hearing loss, the greater the hearing loss the greater the percentage SP. This relationship is shown in figure 3 and has a statistically significant correlation coefficient of 0.57. However, there is no relationship between the length of the history of the vertigo and the percentage SP amplitude. It is further noted that those patients with a fluctuating hearing loss have a smaller percentage SP with mean and standard deviations of 44 ± 12% for the fluctuating group

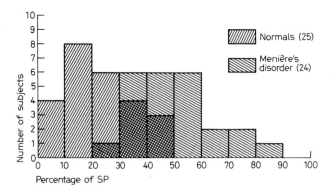

Fig. 2. Percentage SP in normals and Menière's cases.

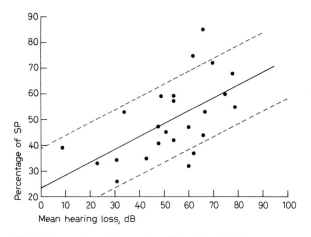

Fig. 3. Comparison of percentage SP and hearing loss.

and $63 \pm 17\%$ for the non-fluctuating group. These means are significantly different at the $p = 0.01$ level. This in turn correlates with a greater hearing loss in those with no fluctuation of hearing.

Figure 4 shows the typical changes in the AP/SP waveform following administration of glycerol, and the mean percentage changes in the SP can be seen in figure 5. The maximum change in the absolute amplitude of the BL/SP from a mean of 1.17 ± 1.28 µV preglycerol to 0.99 ± 0.67 µV postglycerol, was statistically significant at the $p = 0.01$ level. It will be seen that this maximum change occurs at between 30 and 60 min after administration of the glycerol, the mean being 56.4 ± 20.5 minutes.

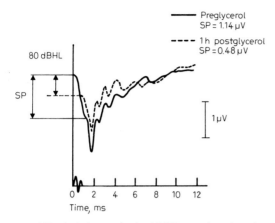

Fig.4. Typical changes in the AP/SP complex after glycerol dehydration.

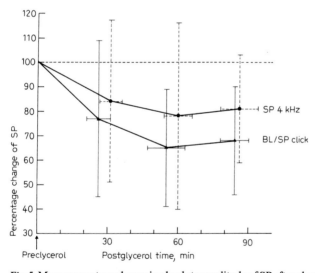

Fig.5. Mean percentage change in absolute amplitude of SP after glycerol dehydration.

In analyzing our results we have given a scoring value to the threshold shift noted on pure tone audiometry recorded during this test. We have recorded two pluses (+ +) for a 10 dB improvement in threshold at three consecutive octave intervals, one plus (+) for a 5 dB improvement and a zero for no change. In some patients a worsening of the auditory threshold of up

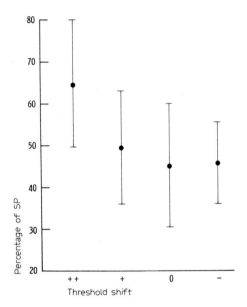

Fig. 6. Comparison of percentage SP and pure tone threshold shift after glycerol dehydration.

to 5 dB was recorded, this being scored minus (–). It can be seen in table II and figure 6, that a greater degree of pure tone threshold shift is found in those patients with a higher percentage SP. However, no relationship was found between the duration of the history and the degree of threshold shift after glycerol dehydration. No significant changes were seen in the amplitude of the CM.

In 9 of the subjects in this series measurement of the serum osmolality was carried out prior to administration of glycerol and at 60 min afterwards. The mean serum osmolality preglycerol administration was 289.1 ± 5.8 mosm/l, that at 60 min postglycerol being 302.6 ± 6.4 mosm/l. This change is statistically significant at the $p = 0.001$ level.

Discussion

In carrying out this investigation we have admitted only patients with classical Menière's symptoms, excluding any 'atypical' cases. All had been fully investigated audiometrically, vestibulometrically and radiologically.

As in our original series [6] no relationship was found between the duration of the history and the degree of hearing loss nor with fluctuation of the hearing loss. In our previous paper we reported finding an enhanced negative SP with a widened AP/SP waveform and a recruiting pattern on the input/output curves. These findings have been confirmed in this larger series. It needs to be stressed that the absolute amplitude of the SP component is not a guide to the presence or absence of hydrops, the range of values overlapping considerably with normal values. There is a mean reduction in the BL/N1 amplitude in Menière's patients, such that the SP component is further enhanced, but again the range of values overlaps considerably with the normal range. *Kumegami* et al. [11] have shown in Menière's patients that the negative SP/AP ratio is greater in those with more severe hearing impairment, the ratio being greater than 0.4 (equivalent to 40% or greater expressed as a percentage SP) in those with the greater loss. This correlates with and supports our own results.

Overall, however, the mean BL/N1 and BL/SP results are statistically significantly different from normals at the p = 0.01 level. The AP/SP waveform is broadened due to the enhanced negative SP but we have not found that measuring its width is of value, as reported by *Gibson* [8], the point at which the response returns to the baseline being difficult to gauge. In view of the wide range of absolute amplitude of both the BL/N1 and BL/SP we feel that the percentage SP becomes an important index. It is of great interest that the percentage SP is greater in those with greater hearing loss.

As in our previous series and as reported by *Moffat* et al. [13], the maximum reduction in the SP occurs at approximately 60 min. Although we have not carried out serial estimations of serum osmolality we have nonetheless found a significant reduction in serum osmolality at 60 min in those patients in whom this measurements was made.

It has been assumed by *Coats* [3] that the increased negative SP/AP ratio or enhancement of the negative SP might be related to endolymphatic hydrops in the cochlea. It is thought that the enhanced negative SP arises as a result of asymmetry produced by endolymphatic hydrops. It is believed by *Gibson* [8], that the SP is a multi-component response arising from various non linear mechanisms within the cochlea, the major component probably resulting from non-linear vibration of the basilar membrane leading to CM being generated disproportionately in one direction. This view receives support from experimental work carried out producing a bias of the basilar membrane by two different methods. *Butler and Honrubia* [2] produced biasing of the basilar membrane using hydrostatic pressure to elevate endolymphatic pressure and *Durrant and Dallos* [5] used low-frequency sound stimulation to produce a bias.

Reduction in the displacement of the basilar membrane would, if these theories are accepted, cause reduction of the enhanced negative SP. *Booth* [1] has suggested that the decrease in the hydrops produces a movement of the basilar membrane towards the scala vestibuli, thus decreasing the asymmetry in the mechano-electrical phenomena associated with the hair cells. His alternative theory is that dehydration could decrease the endolymphatic potential primarily leading to a decreased negative SP. From our own results and from those of *Moffat* et al. [13], supported by the experimental work of *Durrant and Dallos* [5], and *Butler and Honrubia* [2] we believe the former view to be correct, namely that the reduction in the enhanced negative SP is due to a reduced displacement of the basilar membrane following glycerol dehydration.

Acknowledgements

The authors would like to thank *Christine Sills, Nick Setchfield* and all other technical staff of the Evoked Responses Clinic, Medical Physics Department and the Audiology Department, Queen's Medical Centre, University Hospital, Nottingham for their assistance. The illustrations were prepared by the Audio-Visual Department and the manuscript typed by *Angela Beatson*.

References

1   Booth, J.B.: Menière's disease: the selection and assessment of patients for surgery using electrocochleography. Ann. R. Coll. Surg. *62:* 415–425 (1980).
2   Butler, R.A.; Honrubia, V.: Responses of cochlear potentials to changes in hydrostatic pressure. J. acoust. Soc. Am. *35:* 1188–1192 (1963).
3   Coats, A.C.: The summating potential and Menière's disease. Archs Otolar. *107:* 199–208 (1981).
4   Coats, A.C.; Martin, J.L.: Human auditory nerve action potentials and brainstem evoked responses. Archs. Otolar. *103:* 605–622 (1977).
5   Durrant, J.D.; Dallos, P.: Modification of DIF summating potential components by stimulus biasing. J. acoust. Soc. Am. *56:* 562–570 (1974).
6   Gibbin, K.P.; Mason, S.M.; Singh, C.B.: Glycerol dehydration tests in Menière's disorder using extratympanic electrocochleography. Clin. Otolaryngol. *6:* 395–400 (1981).
7   Gibson, W.P.R.; Moffat, D.A.; Ramsden, R.T.: Clinical electrocochleography in the diagnosis and management of Menière's disorder. Audiology *16:* 389–401 (1977).

8 Gibson, W.P.R.: In essentials of clinical electric response audiometry, p.100 (Churchill-Livingstone, Edinburgh 1978).
9 Hallpike, C.S.; Cairns, H.: Observations on the pathology of Menière's syndrome. J. Lar. Otol. *53:* 625–654 (1938).
10 Klockhoff, I.; Lindblom, U.: Endolymphatic hydrops revealed by the glycerol test. Preliminary report. Acta oto-lar. *61:* 459–462 (1966).
11 Kumegami, H.; Nishida, H.; Baba, M.: Electrocochleographic study of Menière's disease. Archs Otolar. *108:* 284–288 (1982).
12 Mason, S.M.; Singh, C.B.; Brown, P.M.: Assessment of non-invasive electrocochleography. J. Lar. Otol. *94:* 707–718 (1980).
13 Moffat, D.A.; Gibson, W.P.R.; Ramsden, R.T.; Morrison, A.W.; Booth, J.B.: Transtympanic electrocochleography during glycerol dehydration. Acta oto-lar. *85:* 158–166 (1978).
14 Schmidt, P.H.; Eggermont, J.T.; Odenthal, D.W.: Study of Menière's disease by electrocochleography. Acta oto-lar. suppl. 316: 75–84 (1974).
15 Yoshie, N.: Diagnostic significance of the electrocochleogram in clinical audiometry. Audiology *12:* 504–539 (1973).

K.P. Gibbin, FRCS, Department of Otolaryngology, Queen's Medical Centre, University Hospital, Nottingham (England)

Adv. Oto-Rhino-Laryng., vol. 31, pp. 208–216 (Karger, Basel 1983)

Possible Utility of Middle Latency Responses in Electric Response Audiometry [1]

H. Davis, S. K. Hirsh, L. L. Turpin

Central Institute for the Deaf, Saint Louis, Mo., USA

The Problem

The most important present application of electric response audiometry is to assess the peripheral auditory function of young, or otherwise difficult-to-test children, in whom hearing impairment is suspected. The general method of response averaging is now familiar, and the use of broad-band clicks to establish the auditory threshold for the frequency range above 1,500 Hz is well established [8,9]. With such stimuli hearing-impaired infants can be identified with confidence [10], and some early decisions as to diagnosis and management can be made, but the best choice of an appropriate hearing aid requires threshold determinations at 1,000 and 500 Hz in addition.

Long tone bursts, which are fully frequency-specific, can be employed if the electric response used as the indicator wave is the slow cortical response [2], and the subject is awake and cooperative. For infants more than 3 months old, however, and for many hyperactive or emotionally disturbed or multiplyhandicapped older children, sleep-like sedation is required to obtain the necessary muscular relaxation. This requirement has led to the use of the brain stem responses, notably the early wave P6 (or Jewett V), as the indicators. The brain stem responses are not affected by sedation. The difficulty now, however, is that at 1,000 and 500 Hz a P6 response is evoked by each sound wave separately. The nerve impulses are no longer synchro-

[1] Supported by a grant from the US Public Health Service, Department of Health Service, Department of Health, Education and Welfare research NS03856 from the National Institute of Neurological and Communicative Disorders and Stroke to Central Institute for the Deaf.

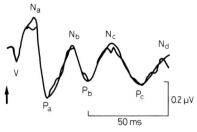

40 dB SL, 500 Hz, 10/s, Σ 2,000

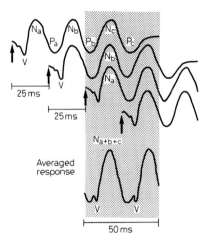

Fig. 1. Middle latency responses (MLR). Top: MLR evoked in a waking adult by 500-Hz tone bursts at 40 dB SL, 10/s. Upward indicates vertex more negative relative to ipsilateral mastoid. Note the overall approximation to a 40-Hz sinusoid, Note also the first negative wave, N_a, with latency about 13 ms. This is the SN10 brain stem response. Bottom: Diagram of superposition of successive MLRs, achieved by stimulating at intervals of 25 ms. This maneuver gives an enhanced averaged response in a relatively brief test period [from ref. 5 by permission].

nized, as they are by high-frequency tone pips or broad-band clicks, and their detectability is reduced considerably [1].

The requirements for frequency selectivity (several sound waves) and sensitivity (good synchronization) are directly opposed. An improvement was made, however, by the introduction of a slower brain stem response, SN10, as the indicator wave at low-frequencies. SN10 integrates brief tone bursts fairly well [3], but it requires the use of an input filter with a lower pass-band. This in turn allows some low-frequency 'noise' from muscles and

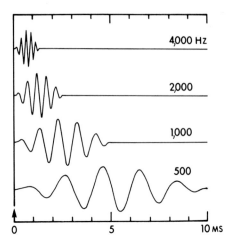

Fig.2. '2-1-2' tone bursts have rise and fall times of 2 periods and a plateau of one period of the modulated frequency. This relation gives them acoustic power spectra that all have the same shape on a logarithmic frequency scale, as shown in figure 3 [from ref. 4].

EEG to enter, and the curving baseline may reduce the sensitivity by as much as 15 dB.

A solution to this difficulty is suggested by *Galambos* et al. [5]. They employ the middle latency response (MLR) with peak latencies from 8 to 100 ms as the indicator. They shorten the collection time substantially by stimulating every 25 ms. The responses partially superimpose on one another, the first wave of the second response on the second wave of the first response, and so on (fig. 1). This is feasible because the MLR is roughly a sinusoid with a frequency of 40 Hz and wavelength (period) of 25 ms. The great advantage is that the relatively slow MLR integrates several waves of a stimulating tone burst, even at 250 Hz. The frequency selectivity at 500 Hz is thus more than adequate for optimal selection of a hearing aid. The sensitivity in waking adults is excellent: identifiable response within 5 dB or less of behavioral threshold. A rather narrow input pass-band is used to eliminate the higher components of muscle 'noise'.

The MLR is reported to be little affected by natural sleep in adults [3, 6, 7]. The remaining question is whether the MLR is equally detectable in sedated children. We regret to report that under our clinical condition of sedation with secobarbital or chloral hydrate the MLR is seriously depressed and offers no advantage for our electric response auditometry.

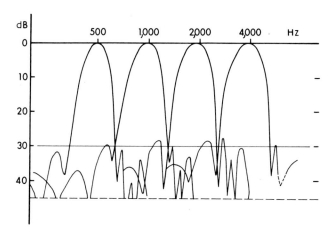

Fig. 3. The envelopes of the acoustic spectra of 2-1-2 tone bursts at 500, 1,000, 2,000 and 4,000 Hz overlap only at 30 dB or more below their peak intensities. This degree of frequency specificity is sufficient for the optimal selection of a hearing aid [from ref. 4].

Method

Our Madsen 2250 ERA unit was modified to permit the Galambos maneuver of stimulating twice, at interstimulus intervals of 25 ms, within a 50-ms window – once at the start and again at the midpoint. The input passband is at 30–80 Hz (3 dB down, with 24 dB/octave slope). The stimuli are very brief tone bursts with 2 periods rise and fall time and 1 period plateau (fig. 2). The envelopes of the acoustic power spectra of these bursts overlap only at more than 30 dB below their peaks (fig. 3). Their itensity has been calibrated relative to the median thresholds of a jury of 8 otologically normal young adults, using nearly the same repetition rate. Trials were run at 1,000 and 500 Hz on several adults, awake or in natural sleep, and also on 19 sedated children at the termination of their routine brain stem response audiometry tests.

EEG-type electrodes were routinely applied with adhesive discs to forehead and to each mastoid, with resistances below 2,000 Ω. Our recording instrument has two input channels. These were usually connected in parallel to forehead and ipsilateral (stimulated ear) mastoid. The opposite mastoid was grounded. One channel was equipped with the MLR filter (30–80 Hz) and the other with our modified 'medium' filter (40–1,100 Hz) [3].

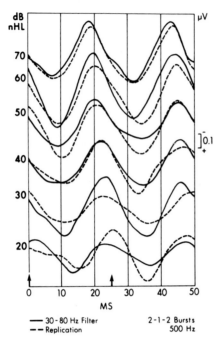

Fig.4. Middle latency responses from a sleepy normal-hearing adult. Stimuli were initiated at times indicated by the arrows, with 25-ms interstimulus intervals. The composite negative (upward) peaks appear about 20 ms later; their latency increases as stimulus intensity is reduced. The long latency is due the low frequency of the pass-band of the input filter: 30–80 Hz with slope of 24 db/octive. Stimuli were 2-1-2 tone bursts at 500 Hz. The dashed lines represent replications of the initial series of trials. $\Sigma = 1,000$ pairs of stimuli.

Results

The responses of an awake but very drowsy adult are shown in figure 4. Only the MLR responses are shown. The 500-Hz stimuli descended in intensity to 20 dB nHL and then the series was repeated. The MLR responses are still large at 20 dB. The subject hears these stimuli clearly at 10 dB nHL but is uncertain at 5 dB. The report of *Galambos* et al. [5] concerning wave form, amplitude and sensitivity in adults is fully confirmed by this and four other less complete tests.

Our efforts to obtain similar results with sedated children are illustrated in figures 5–8. The 'MLR' responses were never of such high voltage as in

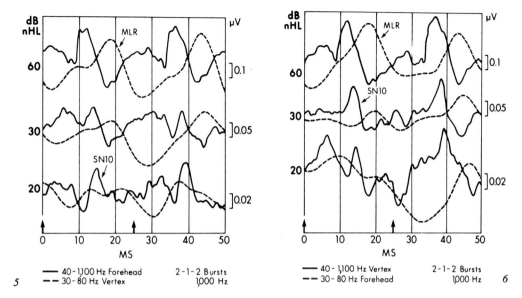

Fig.5,6. The dashed lines represent the superimposed middle latency responses (MLR) to 2-1-2 tone bursts at 1,000 Hz at the hearing levels, indicated at the left. The conditions are comparable to those in figure 4, but this 4-year-old child was sedated with secobarbital and the MLRs are less identifiable than those in figure 4. In a second channel the input filter band was 40–1,100 Hz and the record was smoothed before write-out by three-point averaging. Figures 5 and 6 differ only in the placement of the cranial electrode: MLR from vertex in figure 5 and from forehead, high in the midline, in figure 6. The second channel recorded simultaneously from forehead and vertex, respectively. In the solid-line record with the wider pass-band the SN10 wave appears clearly at 20 dB nHL while in the dashed-line record the composite MLR is not clear. The difference is due to the filters, not to the location of the cranial electrode.

adults. In the figures the responses recorded with the broader pass-band have been smoothed before write-out, by three-point averaging. This enhances the visibility of the SN10 wave, which clearly dominates the response at 60 dB nHL in all cases. The SN10 response appears earlier than the 'MLR' because of the lower frequency of the pass-band of the filter in the MLR channel. The SN10 wave is actually providing most of the voltage of the 'MLR' response, but usually it is more reliably identified in the channel with the wider pass-band, largely because it is preceded by the characteristic dip of the vertex-

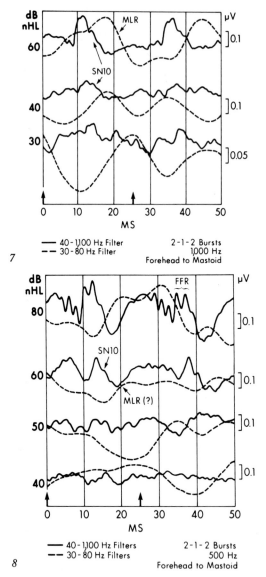

Fig. 7, 8. Middle latency response (MLR) and SN10 recordings were made simultaneously from the same electrodes channels with different input filters. Dashed lines are for
the narrow-band MLR filter, solid for the SN10 filter. The 2-year, 4-month-old child was
sedated with secobarbital. Figure 7 shows responses to 1,000-Hz tone bursts, figure 8 to
500 Hz. The SN10 responses are consistently more recognizable than the composite MLR.
The SN10 threshold at 500 Hz is at 40 dB at 500 Hz by SN10. A trial at 80 dB in the wider-
band channel shows the stimulus artifact followed by the frequency-following response.

positive P6 wave. (Upward is vertex-negative in all of the figures.) Sometimes the 'MLR' and SN10 were equally identifiable, but in well over half of our comparisons the SN10 wave was identified more clearly or at a lower level than the 'MLR'.

In one test we applied electrodes to both vertex and forehead, connected them to channels 1 and 2, respectively, and then reversed the connections and replicated. Figures 5 and 6 show no consistent differences related to the placement of the cranial electrode, but there is the usual advantage for the broader filter. Figures 7 and 8 show tests of another child at 1,000 and at 500 Hz. The responses were no different from 2 children sedated with chloral hydrate instead of our usual secobarbital.

Conclusion

We have tested 19 sedated children more or less completely by this comparative routine and we have found no advantage in the Galambos maneuver. Most of these 19 children had good early BSR responses and thresholds 'within normal limits' (15 dB nHL or better), but the MLRs were rarely identifiable in the Galambos channel at 30 dB nHL. *Galambos* et al. [5] noted such a possibility. The sedation, which we find essential, apparently depresses very severely the components of MLR that arise central to and with longer latency than the brain stem component, SN10. P6 and SN10 are still our two best indicator waves for children. The MLR, as identified by the Galambos maneuver, does not allow us to avoid the audiological compromise between frequency selectivity and sensitivity as we had hoped.

References

1   Davis, H.: Brainstem and other responses in electric response audiometry. Ann. Otol. Rhinol. Lar. *85:* 3–14 (1976).
2   Davis, H.: Principles of electric response audiometry. Ann. Otol. Rhinol. Lar. *85:* suppl. 28, pp. 1–96 (1976).
3   Davis, H.; Hirsh, S.K.: A slow brainstem response for low-frequency audiometry. Audiology *18:* 445–461 (1979).
4   Davis, H.: The infant's audiogram: brainstem electric responses. Auditory development in infancy. Erindale Symp. (Plenum Press, New York, in press).
5   Galambos, R.; Makeig, S.; Talmachoff, P.J.: A 40-Hz auditory potential recorded from the human scalp. Proc. natn. Acad. Sci. USA *78:* 2643–2647 (1981).
6   Mendel, M.I.; Goldstein, R.: Early components of the averaged electroencephalic

response to constant level clicks during all-night sleep. J. Speech Hear. Res. *14:* 829–840 (1971).

7 Mendel, M.I.; Hosic, E.C.; Windman, T.R.; Davis, H.; Hirsh, S.K.; Dinges, D.F.: Audiometric comparison of the middle and late components of the adults auditory evoked potentials awake and asleep. Electroenceph. clin. Neurophysiol. *38:* 27–33 (1974).

8 Picton, T.W.; Woods, D.L.; Baribeau-Braun, J.; Healy, T.: Evoked potential audiometry. J. Otolar. *6:* 90–119 (1977).

9 Picton, T.W.; Stapells, D.R.; Campbell, K.B.: Auditory evoked potentials from the human cochlea and brainstem. J. Otolar. *10:* suppl.9, pp.1–41 (1981).

10 Schulman-Galambos, C.; Galambos, R.: Brainstem evoked response audiometry in newborn hearing screening. Archs Otolar. *105:* 86–90 (1979).

H. Davis, MD, Central Institute for the Deaf, 818 So. Euclid, Saint Louis, MO 63110 (USA)

Adv. Oto-Rhino-Laryng., vol. 31, pp. 217–223 (Karger, Basel 1983)

Clinical Characterization of the Hearing of the Adult British Population[1]

G. G. Browning, A. C. Davis

Medical Research Council, Institute of Hearing Research, Southern General Hospital, Glasgow, Scotland; MRC Institute of Hearing Research, University of Nottingham, Nottingham, England

Introduction

An accurate estimate of the prevalence of hearing disorders in a community is essential to allow forward planning of health service resources and is especially important when there is a large unmet need. Previous studies of the adult British population have suggested a prevalence of 'hearing loss' in the region of 6–8% [1–3] but these studies were conducted many years ago and their methods of data collection could be criticized. In particular the absence of bone conduction measurements to detect conductive disorders and the lack of consideration of noise exposure were significant omissions. With modern developments in microsurgery for conductive disorders and improved methods of hearing aid provision and rehabilitation, up-to-date prevalence data is obviously required.

The aim of the study was to clinically and audiometrically examine a representative sample of the adult British population to enable accurate population predictions to be made. The opportunity was taken to study certain aetiological factors in the causation of a hearing loss. The study has been planned in three phases, the present paper being a preliminary report of the first phase.

[1] This paper was based on a multi-centre study. As such, numerous other investigators took part, At Glasgow, *S. Gatehouse* and *M. E. Lutman;* at Southampton, *P. B. Ashcroft* and *A. R. D. Thornton;* at Cardiff, *J. A. B. Thomas* and *J. J. Miller;* at Nottingham, *R R. A. Coles* and *R. S. Tyler.*

Table I. Prevalence of hearing defects in adult British population in the better (BE) and worse (WE) hearing ear overall and in 4 age categories

	Overall		Age bands							
			15–30 years		31–50 years		51–70 years		70 years +	
	BE	WE	BE	WE	BE	WE	BE	WE	BE	WE
Normal	81	68	96	86	86	72	73	56	27	25
Conductive	6	13	2	12	9	17	6	12	0	3
Mixed	2	5	1	0	0	2	3	8	9	15
Sensorineural	11	14	0.5	1	4	9	17	23	62	56

Method

Sample. In the first phase a random selection of the adult British population aged 17 years and over was taken by sampling from the electoral rolls in Cardiff, Glasgow, Nottingham and Southampton. A total of 11,740 individuals were sent a postal questionnaire enquiring about their hearing status, of which 9,607 (82%) were returned. The sample was stratified on the basis of these replies to allow both dense sampling of certain strata and efficient overall estimates of prevalence. A total of 2,396 individuals, sampled from these strata, were invited to attend for a full otological and audiometric assessment using standardized protocols [4]. Several other investigations were performed but are not reported here. In all 759 (32%) individuals were tested at this second tier. Non-attenders were followed-up on a domiciliary basis. No substantial differences between attenders and non-attenders were found with respect to the measures reported here.

Audiological Definitions. Pure tone thresholds were assessed at 0.25, 0.5, 1, 2, 3, 4, 6 and 8 kHz by air-conduction (AC), and 0.5, 1, 2 kHz by bone-conduction (BC). The mean AC thresholds reported here are for 0.5, 1, 2 and 4 kHz and the mean air-bone (AB) gap for 0.5, 1 and 2 kHz. Auditory function was considered to have a conductive defect if the mean AB gap was greater than 15 dB. A sensorineural defect was defined as a mean AC threshold greater than or equal to 25 dB hearing loss (HL) and no conductive defect was present. A mixed defect was considered present when there was a conductive defect and the mean BC threshold was greater than or equal to 20 dB.

Otological Assessment. A comprehensive otological and general health assessment was made, including blood tests to establish norms and possible

Table II. Effect of age on size of air bone gap in individuals with non-diseased tympanic membranes

Age, years	Male		Female	
	left	right	left	right
17–30	0.55	1.23	0.03	0.12
31–50	2.83	1.60	2.30	3.20
51–70	1.98	0.57	2.66	1.70
70 +	1.05	1.68	1.05	2.82

causes of pathology in extreme cases. For present purposes the most important aspect of the history was the assessment of noise exposure. A noise immission rating (NIR) was calculated from a cumulative noise exposure scale based on extensive questioning concerning military, occupational and recreational exposure. Values of 0, 1, 2 and 3 correspond to lifetime workday exposures of less than 80, 80–90, 91–100 and 101–110 dB (A), respectively.

Statistics. Statistical significance was taken as if ≤ 0.05.

Results

Projecting from the stratified sample to the adult British population, the prevalence of hearing defects is 19% in the better and 32% in the poorer hearing ear (table I). The majority of the defects are of a sensorineural type and become increasingly more common with age. On the other hand, conductive defects do not become commoner with age, the number sampled over the age of 70 years being too small to make the apparent decrease in this age group significant.

The distribution of the a-b gap in the population was as follows; in 5.8% the a-b gap is between 10 and 19 dB, in 1.9% the gap is between 20 and 29 dB and in 1.8% the gap is 30 dB or greater. Only 5% of those with a gap less than 20 dB, 14% with a gap between 20 and 30 dB and 21% of those with a gap greater than 30 dB have had previous surgery. In the 9.5% of the population with a conductive defect, otological examination suggests healed otitis media in 1.9%, inactive chronic otitis media in 1% and active chronic otitis media in 0.6%. In the other 5.6% the tympanic membrane was normal and in most of these the presumptive diagnosis would be otosclerosis.

To assess whether there was an effect of age on the middle ear

Table III. Mean AC thresholds of population without a conductive defect and with noise exposure (NIR = 0) analyzed as to age in 4 bands

Age, years	kHz							
	0.25	0.5	1	2	3	4	6	8
Left Ear								
17–30	7.8	5.2	4.5	4.8	7.3	8.7	13.2	9.1
31–50	9.2	7.2	7.5	7.9	11.7	15.3	21.9	18.7
51–70	13.4	12.1	11.4	15.0	20.2	25.7	37.7	36.3
71 +	19.6	20.6	26.3	35.6	39.9	49.5	62.9	66.9
Right Ear								
17–30	8.3	5.1	5.2	5.4	5.2	7.5	12.4	8.4
31–50	9.8	5.7	7.7	8.4	10.4	14.7	20.3	18.6
51–70	15.2	11.4	12.5	14.4	18.8	25.1	35.8	36.9
71 +	22.6	22.1	26.6	33.6	39.3	47.0	63.0	61.1

conductive mechanism, the size of the AB gap was analyzed excluding individuals with healed or chronic otitis media. In both males and females no progressive effect of age was detected (table II).

Many factors have been analysed as to their effect on AC thresholds and AB gap while controlling for age, sex, socio-economic group and overall noise exposure. To date, eye colour, presence of arcus senilis, family history of hearing loss and proportion of life spent in town as opposed to a rural environment did not appear to correlate with the mean AC thresholds, the mean AB gap or the AC thresholds at any frequency.

Age and noise exposure are by far the most significant factors identified in determining the AC thresholds. There was no significant effect of sex on the mean AC thresholds, though future analysis on a larger sample may show an effect at the higher frequencies. Table III gives the mean AC thresholds of the population without a conductive defect and with no noise exposure (NIR = 0), analysed in 4 age bands. There is a significant effect of age at all frequencies in both ears. The dip at 6 kHz is of similar magnitude in each age group and ear.

Individuals in all age bands if they had had the equivalent of lifetime exposure greater than 90 dB (A) had significantly higher thresholds at 2 kHz and above. The mean magnitude of this effect was 3 dB at 2 kHz, 9.5 dB at 3 kHz, 11 dB at 4 kHz, 9.5 dB at 6 kHz, and 8 dB at 8 kHz.

Of the 80 individuals seen in the clinic with a better ear average AC at 0.5, 1 and 2 kHz greater than 25 dB and no conductive hearing defect, only 17

(21%) were offered and accepted management for their disability. Similarly, of the 130 individuals with a worse ear average greater than 20 dB and a conductive defect, only 21 (16%) were offered and accepted management for their disability.

Discussion

The prevalence of hearing disorders, and in particular sensorineural hearing defects, in the adult British population is considerably higher than previous estimates [1–3]. Though sensorineural defects become more common with age, it is by no means a disease of old age, 9% of those in the age band 31 to 50 years have a mean loss greater than 25 dB in their poorer hearing ear. The majority of individuals with defects have still to seek specialist advice and were they to do so it would represent a considerable work load, even taking into account that only around 20% of those with defects seen in the study accepted management of their hearing defect.

This study has allowed us to examine many aspects of hearing but in particular the effects of aging. No population study to date has measured BC thresholds and controlled for noise exposure. These are obviously important in the assessment of the middle ear conduction mechanism and of air conduction thresholds. It would appear that once individuals with the obvious effects of otitis media have been excluded, there is no progressive effect of age on the middle ear conduction mechanism (table II). This analysis will inevitably have included individuals with otosclerosis but the numbers involved are too small to make any material difference. The fact that the middle ear mechanism does not deteriorate with age is at first surprising. Recent work [5] would suggest that pure tone audiometry is a relatively crude method of assessing middle ear function and it is only when a sizeable defect exists does an AB gap become apparent.

Considering now the effect of age on air conduction thresholds in individuals with no gross conductive defect as measured by pure tone audiometry. Previous population studies [2, 6] have usually controlled for the effect of noise but have perhaps less rigorously excluded individuals with a conductive defect because of the absence of BC thresholds. Despite this, thresholds in the various age bands are surprisingly similar to those measured in 1960–1961, in the non-noise exposed Maaban tribe in a remote area of the Sudan [6]. The difference between age bands is also of the same magnitude as those assessed in 1958 in females in rural Scotland [2].

INTENSITY dB HL (ISO 1964)

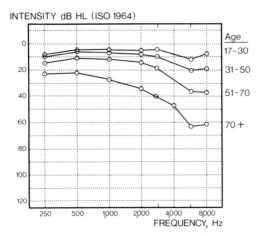

Fig. 1. Mean AC thresholds in right ear of non-noise exposed individuals without a conductive defect, analyzed in four age bands.

When plotted in a standard audiometric form (fig. 1) the relatively small differences in the means for the different agebands become more apparent. These differences may be less than some clinicians have come to expect, perhaps because the Wisconsin state fair survey [7], which included noise exposed individuals and those with a conductive loss, is more widely quoted [8] than the earlier British study [2].

How the clinician utilises this information when investigating a patient is another matter and there is considerable overlap between age groups when the standard deviations from the mean are taken into account. It will be necessary for him to consult a more detailed table than can be drawn up at this stage of our epidemiology study. The clinician will also have to remember that 65% of the variance in thresholds cannot be accounted for by age and noise. In any individual case then other factors, as yet unidentified in population terms, will have to be taken into account. Again in clinical terms it would be wrong to suggest that the effect of aging on hearing, or presbycusis as it is more commonly known, has been clearly defined. It will have to await the identification of these other, as yet unknown, factors. In the interim in those with a sensorineural hearing loss not attributable to noise trauma, it would be more correct, especially if they are under the age of 70 years, to call it an idiopathic sensorineural hearing loss rather than presbycusis.

Noise is by far the most important factor identified in the aetiology of sensorineural defects. The main contributor to this is exposure to industrial and gunfire noise rather than living in a town as opposed to a rural environment. The size of the effect of noise exposure of NIR ≥ 2, which is the equivalent of continuous noise exposure of greater than 90 dB (A), 8 h a day, 5 days per week, 48 weeks per year for 40 years is only in order of 10 dB at 3 to 8 kHz. This is perhaps less than might be expected by clinicians who assess individuals for compensation, but they are most likely to see only those individuals who are most susceptible to noise exposure. Though there is a slightly greater mean effect at 4 kHz, it is only 1.5 dB (HL) greater than the effect at 3 and 6 kHz. These findings will require confirmation in a larger sample of noise exposed individuals before a full reference chart can be compiled for clinical use.

In all analysis thresholds at 6 kHz are marginally greater than those at 4 and 8 kHz. The fact that the size of the difference is the same in all age and noise bands would suggest that the international standard at 6 kHz should be reviewed.

In conclusion, though hearing defects in the adult British community are more prevalent than previously estimated, we are still a long way from defining the aetiology responsible for the large number with sensorineural defects.

References

1   Wilkins, L.: Survey of the prevalence of deafness in the population of England, Scotland and Wales (Central Office of Information, London 1948).
2   Hinchcliffe, R.: The thresholds of hearing as a function of age. Acustica 9: 304–308 (1959).
3   Ward, P.R.; Tucker, A.M.; Tudor, C.A.; Morgan, D.C.: Self-assessment of hearing impairment. Br. J. Audiol. 11: 33–39 (1977).
4   British Society of Audiology: Recommended procedures for pure-tone audiometry using a manually operated instrument. Br. J. Audiol. 15: 213–216 (1981).
5   Gatehouse, S.; Browning, G.G.: A re-examination of the Carhart effect. Br. J. Audiol. (in press).
6   Rosen, S.; Bergman, E.D.; Plester, D.; El-Mofty, A.; Rosen, H.: Presbycusis study of a relatively noise-free population in the Sudan. Trans. Am. Otol. Soc. 50: 135–152 (1964).
7   Glorig, A.; Wheeler, D.; Quiggle, R.; Grings, W.; Summerfield, A.: 1954 Wisconsin state fair hearing survey. Am. Acad. Ophthal. Otolaryngol., Monograph, 1957.
8   Katz, J. (ed): Handbook of Clinical Audiology; 2nd ed., p.430 (Williams and Wilkins, Baltimore 1978).

G.G. Browning, MD, ChB, FRCS (Edin. & Glasgow), Consultant Otologist,
Scottish Section of MCR Institute of Hearing Research, Southern General Hospital,
1345, Govan Road, Glasgow G51 4TF (Scotland)

Adv. Oto-Rhino-Laryng., vol. 31, pp. 224–227 (Karger, Basel 1983)

Ossiculoplasty – Experience with a New Prosthesis

Alastair M. Pettigrew

ENT Department, Stirling Royal Infirmary, Stirling, Scotland

Introduction

Intra-operative time and experience are needed for the surgeon who wishes to acquire the skills of ossiculoplasty using the conventional techniques of remodelled ossicles. The use of ossicles in ossiculoplasty suffers the further disadvantages that ossicles are difficult to procure, or if an autograft is used, the graft has already been affected by disease. This paper described the results of the use of prostheses aimed at overcoming these disadvantages.

Materials and Methods

A piece of cadaveric skull bone measuring approximately 4 × 4 cm was taken and using an electric grinding stone the inner table and the medulla were removed and a lamina of bone 1 mm thick was made from the outer table. Prostheses were cut to the designs shown in figures 1 and 2. The bone was shaped using a 0.75-mm fissure bur and 1.5-mm diameter circular saw bur. A final smooth finish was obtained with a diamond polishing bur. No error greater than 0.25 mm was tolerated for any given dimension. The dimensions of the prostheses were checked against 1-mm ruled graph paper. In addition to the prostheses shown further prostheses were fashioned with a designed length 1 mm longer or shorter than those shown. Thus, a range of sizes was obtained. The prostheses were double-wrapped in transparent paper and sterilized in a high vacuum steam autoclave.

The prosthesis designed to bridge the gap between the handle of the malleus and the head of the stapes (the malleus-stapes bridge) was used on 16 occasions. The malleus-footplate bridge was used on 7 occasions. Many of

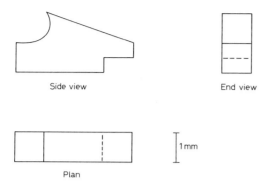

Fig. 1. Malleus-stapes bridge. The curved notch will accept the handle of the malleus, the squared notch will engage the head of the stapes.

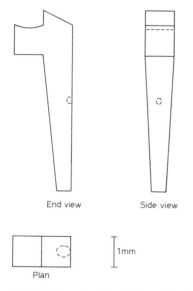

Fig. 2. Malleus-footplate bridge. The short arm of the prosthesis is positioned medial to the handle of the malleus. The long arm contacts the stapes footplate. The small hole in the posterior surface is used in positioning the prosthesis on the footplate.

the ears were affected with tympanosclerosis, some had a perforation, several had an open mastoid or attic cavity. The malleus-stapes bridge was inserted by first engaging the curved notch of the prosthesis on the handle of the malleus. The graft was then swung superiorly to engage the head of the stapes. The two ossicles were pushed slightly apart by the prosthesis. The resulting

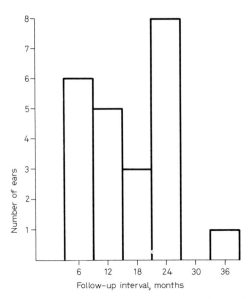

Fig. 3. Histogram. Postoperative follow-up interval.

compression forces maintained the prosthesis in place without any other support. In a similar manner the malleus-footplate bridge was sited, using a hook inserted in the hole in the posterior surface of the bridge.

Results

The follow-up interval for 23 consecutive ossiculoplasties is shown in figure 3. Fourteen of the cases have been followed-up for more than 12 months. Only one prosthesis (a malleus-footplate bridge) was partially extruded and removed, 24 months after insertion. Examination of the removed prosthesis showed that its shape had been conserved apart from some erosion where it was protruding through the tympanic membrane. Histology revealed the prosthesis to be covered with a monolayer of epithelium. There was no new bone growth and no sign of any inflammatory response. One prosthesis was removed because of infection in the ear 15 months postoperatively. The shape of this prosthesis was also well conserved. No new bone was noted. Some white blood cells were present on the surface of the prosthesis. A third prosthesis was removed 4 months after insertion because there had been no gain in hearing. No histology was done on this prosthesis.

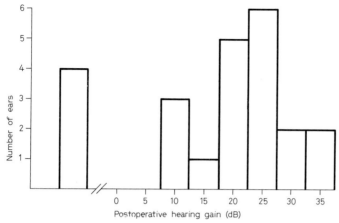

Fig.4. Histogram. Postoperative hearing gain.

The postoperative hearing gain (fig. 4) was obtained by subtracting the postoperative pure tone thresholds (at the longest follow-up interval) averaged for 0.5, 1 and 2 kHz from the preoperative values. In 15 cases there was a hearing gain greater than 15 dB. At the time of follow-up in four ears the hearing was poorer. The fate of three of these grafts is described in the previous paragraph. The fourth was inserted early in the series at the same time as an open mastoidectomy was performed for infected cholesteatoma.

Discussion

The prostheses described can be easily made by a technician in a laboratory after a few hours practice. A ready store of autoclaved prostheses in a range of sizes saves intra-operative time during ossiculoplasty. Their insertion is not difficult. Complications caused by the use of these prostheses are small. Since the other sequelae of chronic otitis media, especially tympanosclerosis, are often present in addition to ossicular discontinuity, ideal hearing results cannot be expected by the use of any prosthesis which aims solely at restoring the continuity of the chain. Most patients in whom the prostheses have been used, however, are aware of an improvement in their hearing.

A.M. Pettigrew, FRCS, ENT Department, Stirling Royal Infirmary,
Stirling, FK8 2AU (Scotland)

Adv. Oto-Rhino-Laryng., vol. 31, pp. 228–239 (Karger, Basel 1983)

Common Complications following Removal of Vestibular Schwannoma

William W. Montgomery

Massachusetts Eye and Ear Infirmary, Boston, Harvard Medical School,
Boston, Mass., USA

There are numerous complications following the translabyrinthine, middle fossa, and suboccipital approaches to the internal auditory meatus and cerebellopontine angle to resect the vestibular schwannoma. This paper will deal only with the most troublesome and most common complications which result from facial nerve paralysis and cerebrospinal fluid oto-rhinorrhea.

Facial Nerve Paralysis – Lagophthalmos

If facial nerve paralysis does not have onset until several days following surgery, then one can be almost certain that return of function will occur within weeks or a few months. In these cases, for the most part, the lagophthalmos can be treated providing the patient has a reasonable Bell's phenomenon by conservative management consisting of drops, ointment and measures to protect the cornea.

Artificial tears such as methocellulose (1%) or polyvinyl alcohol (1.5%) can be used every 1 or 2 h during the waking hours. Ointment may also be necessary during the waking hours if the drops are not effective in controlling any eye discomfort, if the patient has a poor Bell's phenomenon, or if there is decreased sensation of the cornea.

Ointment (erythromycin 0.5%) is used at bedtime and the lids are kept closed using clear plastic tape (Transpore). Tincture of Benzoin is applied to the skin and allowed to dry prior to taping. When the tape is properly applied, vision should be obscured. This is much preferable to an eyepatch which can cause injury to the cornea if the lids remain open under the patch.

Another technique for taping which can be quite successful and cosmetically pleasing is the use of half inch surgical skin closure tape such as Steri-strips. One 12-mm inch Steri-strip is applied to the center of the lower lid and then pulled laterally and superiorly thus elevating the lower lid. In some patients this technique can be used with sufficient success so that a lateral tarsorraphy is not necessary.

The patient should wear protective glasses such as the wrap around type or goggles. This is especially true when the patient is out of doors and exposed to wind and dust. Goggles are, of course, the most satisfactory for they form a humidity chamber. Clear plastic eye shields with tape around the periphery can be purchased. This device can be left in place constantly and eliminates the necessity for drops and ointment. The patient can see through the clear plastic and cosmesis is quite satisfactory.

A lateral tarsorraphy procedure is indicated when recovery of facial nerve function is not expected for at least 3 months as well as in patients who have a poor Bell's phenomenon, severe lagophthalmos, or a dry eye. Ophthalmology consultation and a medial and lateral tarsorraphy is indicated in patients who have anesthesia of the cornea along with lagophthalmos.

There are numerous other methods for the treatment of lagophthalmos and protection of the cornea in patients with paralysis of the orbicularis oculi muscle, none of which in my experience, are as satisfactory as the techniques mentioned above.

Technique of Lateral Tarsorraphy

The technique of the lateral tarsorraphy operation is herewith presented. When properly executed this is the most effective method for protecting the eye, the cosmetic result is not too objectionable, and the procedure can be easily reversed when either the facial nerve function has returned or the hypoglossal facial anastomosis becomes effective (approximately 6 months). The lateral tarsorraphy can be performed as an office procedure, in a minor operating room, in the operating room, or even at the patient's bedside. The necessary equipment and solutions: Betadine solution (povidone iodine); Proparacaine 1%; 1% xylocaine with epinephrine; normal saline solution; 0.5% erythromycin ophthalmic ointment; No. 5–0 polyethylene suture (Dermalene); No. 90 polyethylene or 0.762 × 1.65 mm silicone tubing; No. 28 or 30 short hypodermic needle; 2-ml syringe; No. 15 Bard Parker blade;

knife handle; needle holder; fixation forceps; fine-toothed forceps; scissors –
straight iris, curved iris, suture.

Procedure. The lateral tarsorraphy procedure is performed with the
patient in the supine position and the surgeon seated at his side. Preoperative
medication is not indicated. One drop of topical anesthesia such as proparacaine 1% is placed in the eye. This is repeated in 5 min. The eye is
prepared with Betadine solution which is allowed to dry before applying a
plastic drape. The lid margins are anesthetized using 1% xylocaine with
epinephrine. The injection of local anesthesia is begun at a point 5 mm lateral
to the lateral canthus. The lateral half of each lid margin is infiltrated by
gradually pushing the local anesthetic agent ahead of the needle point
(fig. 1a).

The upper lid margin is grasped with a fine-toothed forceps approximately 2 or 3 mm from the lateral canthus. The epithelium under the forceps
is cut using a straight or curved iris scissors, thus exposing the tarsal plate.
The lateral canthal angle must not be violated so that the tarsorrhaphy can be
reversed at a later date if the orbicularis function returns. A strip of
epithelium about 6 mm in length is cut from the upper lid margin, and is
temporarily left intact medially.

The epithelium over the lower lid margin is grasped a millimeter or two
more medially in comparison with the upper lid and a strip of equal length is
elevated and left intact medially.

The two strips of epithelium ('handles') are grasped with forceps and
pulled so as to bring the medial aspect of both denuded areas in contact. This
is a test to determine whether or not a sufficient length of lower lid has been
resected so as to properly correct the lagophthalmos. Once this correct
amount has been determined the 'handles' are removed. A monofilament 5-0
polyethylene suture supported by either plastic or rubber tubing (No. 90
polyethylene tubing or silicone rubber tubing) is best suited for this operation. If the tubing is not used, the suture will be eroded in a day or two by
the levator palpebrae superioris which is innervated by the third cranial
nerve. The suture is placed as follows (fig. 1b): (a) down through the upper
tarsal plate at the lateral margin of the denuded strip, (b) through the lateral
aspect of the lower lid margin, (c) threaded through a 6-mm piece of tubing,
(d) up through the medial aspect of the lower denuded lid margin, (e) through
the medial aspect of the upper denuded lid margin, (f) through a second 6-
mm piece of tubing, and (g) is tied at the upper lateral aspect of the repair
with multiple knots (fig. 1c).

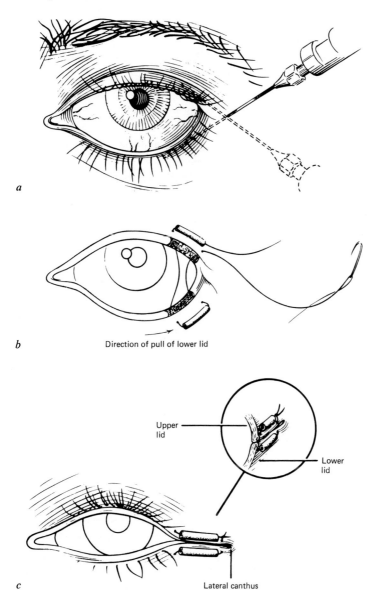

a

b   Direction of pull of lower lid

Upper
lid

Lower
lid

c   Lateral canthus

Fig. 1. a The injection of local anesthesia is begun at a point 5 mm lateral to the lateral canthus. The lateral half of each lid margin is infiltrated by gradually pushing the local anesthetic solution ahead of the needle point. *b* Technique for applying sutures with lateral tarsorraphy operation. *c* The suture is tied at the upper outer aspect of the repair with multiple knots.

At the end of the repair, blood is irrigated from the conjunctival sac with normal saline solution. Once this has been accomplished, 0.5% erythromycin ointment is placed in the eye and the lids are closed. An eye patch is secured in place with one strip of plastic adhesive tape.

The patient is instructed to apply erythromycin ointment in the eye for 2 days, keeping the eye patched at all times. Thereafter, the patient should require no therapy unless there is an insufficient production of tears, in which case it is necessary to continue the usage of artificial tears every 2 h and erythromycin ointment at bedtime. It is most important to leave the tarsorraphy sutures in place for at least 2 weeks; otherwise the attachment between the lids will become separated or stretched.

Facial Nerve Paralysis – Rehabilitation

Numerous surgical procedures have been devised for rehabilitation following facial nerve paralysis when function is not recovered. There are methods for suspending the paralyzed face using fascia, tendons, and synthetic material. Muscle transfer techniques include the masseter and temporalis muscle transfer procedures. Face-lifting procedures have been accomplished in conjunction with the above. Nasolabial fold and forehead skin wrinkles can be constructed by excising skin and repairing the linear defect with inverting suture techniques. As a general rule, these procedures are not too gratifying, offering a very limited cosmetic and little or no functional improvement.

Hypoglossal Facial Anastomosis. In my experience, the hypoglossal facial anastomosis is by far superior to any other rehabilitative technique for relief of signs or symptoms associated with facial paralysis. The operation is not difficult. I have obtained a fairly good return of function even in patients who have had complete paralysis for 6 years prior to the operation. Facial symmetry can be expected in 5–6 months following hypoglossal facial anastomosis. The contraction of facial musculature is first seen as the patient strongly presses the end of this tongue against his hard palate. In a period of about 6 months, the patient can educate the facial musculature to contract voluntarily. The disability resulting from sectioning the hypoglossal nerve is not severe. The patient may have some difficulty with chewing and speech articulation for a few weeks after the operation. This is, as a general rule, not a significant problem.

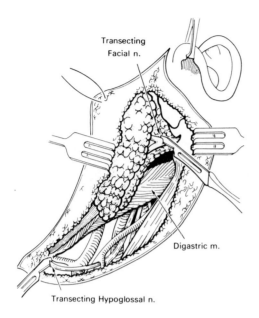

Transecting
Facial n.

Digastric m.

Transecting Hypoglossal n.

Fig. 2. The facial nerve has been dissected, exposing the trunk at its exit from the stylomastoid foramen. It is necessary to elevate the superficial lobe of the parotid gland for proper exposure. The facial nerve is transected as close as is possible to the stylomastoid foramen. The hypoglossal nerve is dissected forward beneath the inferior border of the posterior belly of the digastric muscle. It is transected at the anterior aspect of this muscle.

The incision is similar to that for a parotidectomy beginning in the preauricular crease, with the exception that it extends inferiorly midway between the anterior border of the sternocleidomastoid and the submandibular triangle.

The facial nerve is identified as with a parotidectomy operation using the pointer of the conchal cartilage, mastoid process, styloid process, and tympanomastoid suture as guides. It is transected as it exits from the stylomastoid foramen (fig. 2). The hypoglossal, vagus, and spinal accessory nerves are identified in the angle between the posterior belly of the digastric muscle and the sternocleidomastoid muscle, using the tubercle of the transverse process of the atlas as a guide. The hypoglossal nerve is followed inferiorly and the descendens hypoglossi is identified and sectioned. From here the hypoglossal nerve is dissected forward beneath the inferior border of the posterior belly of the digastric muscle. It is transected at the anterior aspect

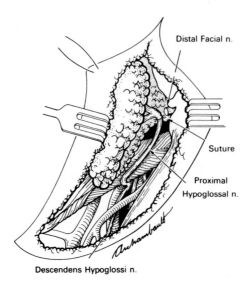

Distal Facial n.

Suture

Proximal
Hypoglossal n.

Descendens Hypoglossi n.

Fig. 3. The hypoglossal nerve is reflected superiorly over the posterior belly of the digastric muscle in order to approach the distal end of the facial nerve. Anastomosis is accomplished as described in the text.

of this muscle. It is reflected superiorly over the posterior belly of the digastric muscle in order to approach the distal end of the facial nerve. The equipment necessary for a nerve anastomosis consists of a surgical microscope or loupes, 2 jeweler's forceps, and No. 9–0 or No. 10–0 black silk or monofilament nylon suture material. The anastomosis is accomplished with 5–6 sutures carefully placed in the nerve sheath (fig. 3). Closed suction drainage is preferred to open drainage.

As soon as facial nerve function has returned, considerable time must be spent with the patient instructing him to develop the newly gained neuromuscular function. The best method to accomplish this end is for the patient to play 'monkey before the mirror'. The tongue is pressed firmly against the hard palate which stimulates the facial muscles to function. This exercise should be carried out at least 15 min twice a day. After many months the patient (especially younger people) can overcome some degree of synkinesis, i.e. avoiding eye closure while smiling. Reinnervation of the frontalis muscle is sometimes unsatisfactory and an eyebrow elevation procedure is necessary for cosmesis.

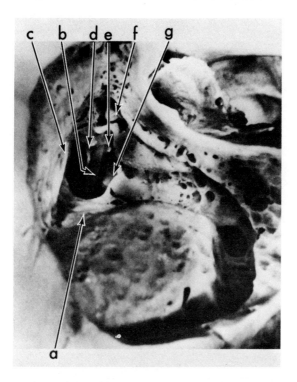

Fig.4. The posterior lip of the internal auditory meatus and a portion of the bony plate anterior to the posterior fossa dura has been resected. The intracanalicular portion of the tumor has been resected with the vestibular and cochlear nerves. The facial nerve remains intact. The cells superior to the internal auditory meatus extend to the petrous tip. The transverse crest of the internal auditory meatus can be visualized. Sufficient exposure for resectioning the intracranial portion of the tumor has been accomplished. Additional posterior fossa dura plate can be removed if necessary. a = posterior fossa dura plate; b = facial nerve; C = position of superior petrosal sinus; d = superior wall of internal auditory canal; e = transverse crest of internal auditory meatus; f = head of malleus; g = inferior wall of internal auditory canal [2].

Cerebrospinal Fluid Otorhinorrhea

Cerebrospinal fluid otorhinorrhea is uncommon following the middle cranial fossa approach to the internal auditory canal and can be prevented by obliterating the labyrinthectomy and mastoidectomy defects with adipose subcutaneous tissue when the translabyrinthine operation is employed. There is significant incidence of cerebrospinal fluid otorhinorrhea following the

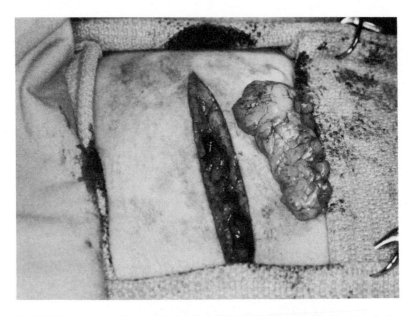

Fig. 5. Subcutaneous adipose tissue is obtained by way of a horizontal incision in the lower left abdomen. The autograft should not be traumatized.

suboccipital approach to the cerebellopontine angle and internal auditory canal by this route (18%). This complication could be avoided by performing a mastoidectomy and exposing the posterior fossa dura prior to the suboccipital operation and obliterating the defect with adipose tissue. Exposure of the posterior fossa dura would facilitate dissection of the internal auditory canal by way of the suboccipital route. So far, we have not employed this technique.

Following the translabyrinthine operation there is a sizeable defect which includes the posterior aspect of the internal auditory canal as well as the adjacent posterior fossa dura (fig. 4). The field is observed carefully for at least 15 min to make certain that the cerebrospinal fluid is clear and that there is no bleeding. Subcutaneous adipose tissue can be taken from the left lower abdomen using a horizontal incision (fig. 5). The adipose tissue should be obtained immediately prior to its implantation. Studying the fate of adipose tissue implants in a bony cavity clearly demonstrates that both trauma to and drying of the adipose tissue prior to its implantation are detrimental to its survival. If the adipose tissue is obtained atraumatically and immediately prior to its implantation, revascularization of adipose tissue

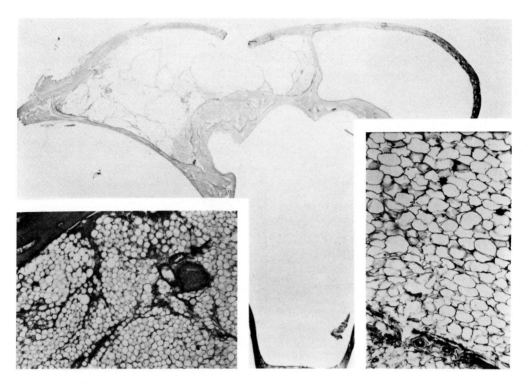

Fig.6. The left frontal sinus was implanted with adipose tissue after the entire mucous membrane lining and inner cortical lining of the sinus had been removed. The insert on the right is a photomicrograph of subcutaneous adipose tissue at the time it was taken from the abdominal wall. The insert on the left is a photomicrograph of the 1 week adipose tissue implant in the left frontal sinus. For the most part this implant will survive as adipose tissue with varying amounts of change to fibrous tissue. Either way a complete and effective obliteration is accomplished.

from ingrowth of blood vessels from the bone will occur during the first few days following implantation (fig. 6, 7).

A small piece of adipose tissue, 1–2 cm in diameter, is placed in the translabyrinthine defect. A second small piece is inserted in the mastoid antrum and the remainder of the mastoidectomy defect is obliterated using a third larger piece of adipose tissue. The incision is closed in layers and not drained. A gauze wick is inserted into the external auditory canal and a mastoidectomy type dressing is applied.

Cerebrospinal fluid otorhinorrhea following the suboccipital operation can occur immediately following the operation or be delayed by a week or

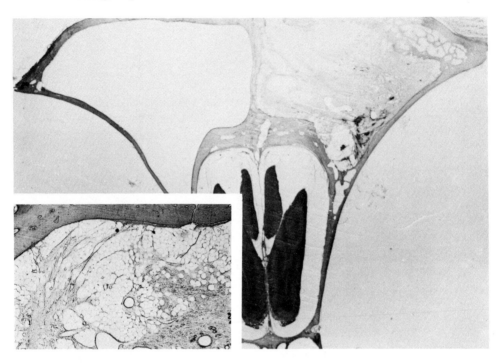

Fig. 7. The left frontal sinus was implanted with adipose tissue as has been described. The specimen was obtained after perfusion with formalin at the end of 1 year. The bony cavity remains well obliterated. Approximately 75% of the adipose tissue survived (insert).

two. As a general rule, there is no question concerning this diagnosis for the rhinorrhea is profuse and on the same side as the operation. It increases with any straining or as the patient leans forward. It is best to repair the dural defect using the technique mentioned above within a week or two for there is a real possibility of the patient being further complicated by meningitis.

A lumbar puncture is accomplished several hours or the evening prior to surgery and the spinal fluid pressure recorded. In some patients with a large dural defect, the spinal fluid pressure is quite low making a spinal tap in the recumbent position quite difficult. In these cases the tap is carried out with the patient in the sitting position. Fluorescein (0.5 ml of 10% solution) is diluted with at least 10 ml of spinal fluid and slowly injected intrathecally. In cases in which 10 ml of spinal fluid cannot be obtained because of low spinal fluid pressure, the fluoroscein can be diluted with normal saline solution.

Following this the diagnosis is confirmed by observing fluorescein-stained rhinorrhea and observing the fluorescence behind the tympanic membrane.

A complete mastoidectomy is accomplished. Usually the dural defect leads to the cells around the sigmoid sinus. Occasionally, however, the route of spinal fluid leakage will be by way of the internal auditory canal and the petrous cells. In such cases further dissection is necessary. Following identification of the site of leakage, the complex is obliterated with adipose tissue as described above.

References

1   Montgomery, W.W.: Surgery of the upper respiratory system; 2nd ed., vol.I, chap.5, 11 (Lea & Febiger, Philadelphia 1979).
2   Montgomery, W.W.: Surgery of the upper respiratory system, vol.II, chap.3 (Lea & Febiger, Philadelphia 1973).
3   Montgomery, W.W.: The fate of adipose implants in a bony cavity. Laryngoscope, St.Louis 74: 816–827 (1964).
4   Peer, L.A.: The neglected free fat graft. Plastic reconstr. Surg. 18: 233–250 (1956).
5   Montgomery, W.W.; Van Orman, P.G.: The inhibitory effect of adipose tissue on osteogenesis. Ann. Otol. Rhinol. Lar. 76: 988–998 (1967).

W.W. Montgomery, MD, Massachusetts Eye and Ear Infirmary,
Boston, MA 02114 (USA)

Adv. Oto-Rhino-Laryng., vol. 31, pp. 240–246 (Karger, Basel 1983)

Fatality Following the Use of Low Molecular Weight Dextran in the Treatment of Sudden Deafness [1]

George M. Zaytoun[a], *Harold F. Schuknecht*[b], *Howard S. Farmer*[c]

[a] Department of Otolaryngology, American University of Beirut, Lebanon;
[b] Department of Otology and Laryngology, Harvard Medical School, and
Department of Otolaryngology, Massachusetts Eye and Ear Infirmary,
Boston, Mass., USA; [c] Section of Otolaryngology, Rutgers Medical School,
Piscataway, N.J., USA

Rheomacrodex®, a low molecular weight dextran (LMWD), has an average molecular weight of 40,000 and is administered intravenously in a 10% solution in either normal saline or in 5% dextrose in water. The principal uses for the drug have been as adjunctive therapy in shock and as a hemodiluent in the extracorporeal circulation [1–3].

A review of the literature reveals that infusion therapy with Rheomacrodex has also been advocated for the treatment of sudden deafness, noise deafness and Menière's disease.

In this paper we will report an unfortunate result of treatment of sudden deafness with this drug.

Case Report

History

The patient was a 46-year-old healthy white male who while sitting quietly at the table sipping his evening coffee suddenly felt slightly lightheaded and vertiginous. He noticed a 'squishy' feeling in both ears. Seconds later, he felt the loss of hearing and the onset of tinnitus in both ears, worse in the right. The next morning he felt that his left ear was completely normal, but he still had hearing loss and tinnitus in the right ear. The vertigo had completely disappeared. There were no symptoms to suggest an upper respiratory infection, viral infection or central nervous system disturbance.

5 days after the onset of hearing loss he presented himself for otologic consultation at which time ENT examination proved to be entirely unremarkable. Otoneurological

[1] This work was supported by NINCDS Grant No. 5 R01 NS05881-16.

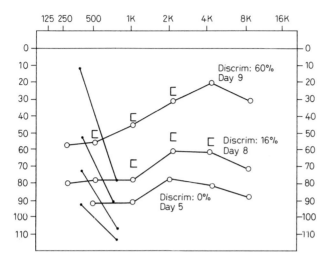

Fig. 1. Sequential audiograms showing hearing improvement on days 5, 8, and 9, following the onset of sudden deafness. On day 11 his hearing was subjectively normal. The alternate binaural loudness balance test was performed on day 8. Rheomacrodex was given from day 6 to 13 following the onset of hearing loss.

examination including fistula test was unremarkable. Audiogram revealed normal hearing in the left ear. In the right ear he demonstrated a 75- to 90-dB sensorineural hearing loss for all frequencies with a 90-dB speech reception threshold and nonmeasurable speech discrimination. Tympanogram was normal. A stapedius reflex was elicited in the left ear with 105 dB stimulation in the right ear suggesting the presence of recruitment in the right ear. No stapedius reflex could be elicited from the right ear with either ipsilateral or contralateral stimulation. The following day he was admitted to the hospital for further studies and therapy. Medical workup including routine CBC and urinalysis, VDRL, electrolyte determinations, thyroid evaluation, SMA 12 and evaluation of serum lipids was entirely unremarkable. The erythrocyte sedimentation rate was 6 mm/h. Stool guaiac was negative. Medical consultation was obtained. Electrocardiogram showed sinus bradycardia. Chest X-ray was normal. Neurological consultation was obtained. A spinal tap indicated clear fluid with no cells. Chemistries were normal. Brain scan was normal. Skull X-rays and polytome examinations of the temporal bones were unremarkable.

The patient was placed on bedrest and given Hygroton 50 mg/day. The next morning a continuous infusion of Rheomacrodex (Dextran LMD40) in 5% glucose was started and given at a rate of 1,500 ml per 24 h. He also received inhalation treatments (95% O_2/5% CO_2) for 20 min 4 times per day. Intermittent infusions of histamine phosphate (2.75 mg in 250 ml of 5% dextrose in water) were given 3 times a day. The rate of flow was adjusted so that the patient experienced a flush and throbbing pulse, but was kept slow enough so that headache was bearable. The histamine treatments frequently led to vomiting causing him to refuse meals, and often they had to be suspended. Audiograms showed a steady improvement in pure tone thresholds and speech discrimination scores (fig. 1).

Because of improvement in hearing the treatment team was encouraged to persist with the Rheomacrodex infusions. On day 11 following the onset of deafness (fifth day of treatment) he felt that the hearing had returned to normal in his right ear, but because of general malaise, he refused to be transported to the audiological facility for a hearing test. He received the last of the Rheomacrodex infusions on day 13. On the following day renal failure was diagnosed. Sodium determination at that time was 106 mEq/l. Urine output was only 10 ml per day. He became visibly swollen, short of breath and demonstrated moist rales throughout his lung fields. His condition continued to deteriorate. On day 17 peritoneal dialysis was begun. As he was receiving the third liter of fluid, his breathing became increasingly shallow and he went into respiratory arrest followed shortly by cardiac arrest. He was successfully resuscitated and the dialysis was continued. Several hours later, as he was nearing the completion of the infusion of another 2 liters of fluid, he again went into respiratory and cardiac arrest. Again he was resuscitated, but he never fully regained consciousness, developed generalized seizures, and went into a downhill course which ended with his death 27 days following the onset of his hearing loss. The temporal bones were removed at the time of autopsy 11 h after death. Autopsy revealed a perforated duodenal ulcer and peritonitis, lobular pneumonia, osmotic nephrosis, and ischemic degeneration of the brain. The temporal bones were prepared for histological study by the standard method of decalcification, imbedding in celloidin, serial sectioning, staining and mounting on glass slides.

Histopathology

Right Ear. There are no unusual findings in the middle ear. The stapes footplate and round window membrane are intact. There is moderately severe postmortem autolysis of the membranous labyrinth; however, all cytological elements can be identified (fig. 2). The cochlear hair cell population is normal. The spiral ligament shows atrophic changes consistent with age. The cochlear neuronal population appears to be normal. The saccular and utricular maculae and the cristae appear normal. The vein at the cochlear aqueduct and the cochlear acqueduct are both patent and show no abnormalities. All arteries and veins appear normal. The endolymphatic and perilymphatic spaces are free of precipitate.

Left Ear. The findings in the left ear are essentially the same as in the right ear.
In summary, histological studies show normal sensory, neural, and vascular structures in both ears. There is no evidence of inflammatory reaction or precipitate in the fluid spaces. The oval and round windows show no evidence of fistulization.

Discussion

Adverse reactions are more common with higher molecular weight dextrans, but may also occur with those of lower molecular weight. The following complications have been observed: (1) hypercoagulability with formation of microthrombi [4], (2) alteration of the structure and function of factor VIII (antihemophilic factor) resulting in bleeding [4], (3) anaphylactoid reactions [4, 5], and (4) renal complications [5, 6].

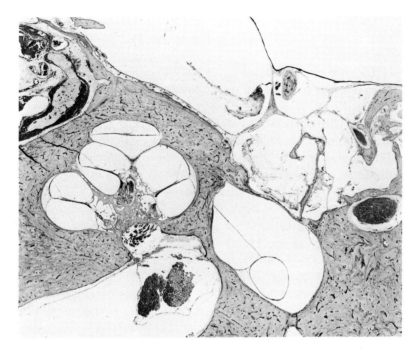

Fig. 2. Mid-modiolar section of right temporal bone showing normal middle and inner ear structures (day 27 following the onset of deafness).

Anuria may follow a brief episode of oliguria. This complication is most often seen following prolonged treatment (e. g. 10 days) and is dose-related [5]. Renal pathology is characterized by massive edema of proximal tubular cells [5].

Among the precautions recommended by the manufacturer of Rheomacrodex 40 are the following: (1) Assess renal function before and during therapy. (2) Assess hemodynamics and bleeding factors before and during therapy. (3) Should therapy continue beyond 24 h, total daily dosage should not exceed 10 ml/kg body weight and therapy should not continue beyond 5 days.

Most of the reports that deal with the clinical use of LMWD in the treatment of acute noise deafness have claimed satisfactory results [7–11]. The rationale is based on the assumption that a vascular disturbance is the cause for the sensorineural hearing loss resulting from exposure to intense noise or blast.

LMWD and Papaverin were used by *Rossberg and Kruger* [12] in the treatment of 50 patients with Menière's disease. They reported success in the preservation of hearing.

The rationale behind the use of LMWD in the treatment of idiopathic sudden deafness is also based on the concept of an underlying vascular disturbance [13,14]. Several observations support this view. *Bomholt* et al. [15] found low fibrinolytic activity and increased platelet aggregability in 18 patients with idiopathic sudden deafness. *Zajtchuk* et al. [16] found hypercoagulability, as evidenced by increased prothrombin consumption, in patients suffering from sudden sensorineural hearing loss. *Ruben* et al. [17] described sudden hearing loss in patients with macroglobulinenemia and blood hyperviscosity. In clinical trials, however, *Knothe* [18], *Giger* [13], *Russolo* [19], and *Strauss and Kunkel* [20] were not convinced of the therapeutic usefulness of LMWD in restoring hearing following idiopathic sudden hearing loss.

Mattox and Simmons [21] found the incidence of spontaneous recovery of hearing from sudden deafness to be 65%. *Wilson* et al. [22] found the recovery rate for severe hearing losses, such as in our patient, to be 24%.

While the recovery of hearing in the case cannot be unequivocally attributed to treatment with Rheomacrodex 40, one of us (H.S.F.) has the clinical impression that the drug is of benefit and continues to use it in a carefully monitored regime.

The pathology of sudden deafness has been described in several papers by *Schuknecht* and his colleagues [23–25]. Viruses have also been implicated as a cause of sudden deafness by *Veltri* et al. [26]. In most cases of permanent hearing loss the principal pathology is a loss of sensory cells in the cochlea. Although these findings are more compatible with a viral etiology than a vascular disturbance, the etiologic mechanism for sudden deafness has not been firmly established.

Conclusions

The experiences with this case of sudden deafness lead to several conclusions: (1) An ear suffering a severe sudden hearing loss showed no pathological changes 27 days following the onset of deafness and 16 days following the return of hearing to normal. (2) This report provides no evidence to support or denigrate the value of LMWD in the treatment of sudden deafness, nor does it provide information on the etiology of sudden deafness, (3) If LMWD (Rheomacrodex 40) is to be used for the treatment of sudden deafness or any other otological disorder, it should be administered with strict observance of recommended precautions and dosages.

References

1   Hint, H.: Propriétés physiochimiques du dextran. Anesth. Analg. Réanim. *33:* 487–493 (1976).
2   Messmer, K.: Les effets de l'hemodilution sur les propriétés rhéologiques du sang et sur l'oxygenation du sang. Anesth. Analg. Réanim. *33:* 502–509 (1976).
3   Bergentz, S.E.: Effect du dextran sur les phénomènes de coagulation et d'hémostase. Anesth. Analg. Réanim. *33:* 603–608 (1976).
4   Laxenaire, M.C.; et al.: Incidents et accidents observés lors de l'utilisation des macro-molecules. Ancsth. Analg. Réanim. *33:* 19–30 (1976).
5   Hint, H.: Effets indésirables des macromolecules. Anesth. Analg. Réanim. *33/4:* 31–36 (1976).
6   Matheson, N.: Renal failure after administration of Dextran 40. SGO *131:* 661 (1970).
7   Kellerhals, B.; Hippert, F.; Pfaltz, C.R.: Treatment of acute acoustic trauma with low molecular weight dextran. Practica oto-rhino-lar. *33:* 260–264 (1971).
8   Kellerhals, B.: Acoustic trauma and cochlear microcirculation. An experimental and clinical study on pathogenesis and treatment of inner ear lesions after acute noise exposure. Adv. Oto-Rhino-Laryng., No.18, pp.91–168 (Karger, Basel 1972).
9   Jakobs, P.; Martin, G.: The treatment of acute acoustic trauma (blast injury) with Dextran 40. HNO *25:* 349–352 (1977).
10 Martin, G.; Jacobs, P.: Clinical comparison of Dextran 40 and Xantinol Nicotinate in the treatment of sudden deafness caused by shock waves. Laryngol. Otol. Rhinol. *56:* 860–863 (1971).
11 Eibach, H.; Borger, U.: Acute acoustic trauma. The therapeutic effect of Bencyclan in a controlled clinical trial. HNO *27:* 170–175 (1979).
12 Rossberg, G.; Kruger, E.C.: The treatment of sudden deafness and Menière's disease with Papaverin and low molecular weight dextran. Laryng. Rhinol. Otol. *56:* 160–166 (1977).
13 Giger, H.L.: Therapy of sudden deafness with O_2-CO_2 inhalation. HNO *27:* 10–19 (1979).
14 Jaffe, B.F.: Sudden deafness. An otologic emergency. Archs Otolar. *86:* 55–60 (1967).
15 Bomholt, A.; Bak-Pedersen, K.; Gormsen, J.: Fibrinolytic activity with sudden sensorineural hearing loss. Acta oto-lar. suppl.360, pp.184–186 (1979).
16 Zajtchuk, J.; Falor, W.H.; Rhodes, M.F.: Hypercoagulability as a cause of sudden neurosensory hearing loss. Otolaryngol. Head Neck Surg. *87:* 268–273 (1979).
17 Ruben, R.J.; Distenfeld, A.; Berg, P.; Carr, R.: Sudden sequential deafness as the presenting symptom of macroglobulinemia. J. Am. med. Ass. *9:* 1364–1365 (1969).
18 Knothe, J.: Die Horsturzbehandlung als klinisches Problem. HNO *27:* 159–164 (1979).
19 Russolo, M.: Acute idiopathic auditory failure; prognosis. A review of 65 cases. Audiology *19:* 422–433 (1980).

20 Strauss, P.; Kunkel, A.: Sympathetic trunk treatment or infusion therapy in cases of sudden deafness. Laryngol. Rhinol. Otol. *56:* 366–371 (1977).

21 Mattox, D.E.; Simmons, F.B.: Natural history of sudden sensorineural hearing loss. Ann. Otol. Rhinol. Lar. *86:* 463–480 (1977).

22 Wilson, W.; Byl, F.; Laird, N.: The efficacy of steroids in the treatment of idiopathic sudden hearing loss (in press).

23 Schuknecht, H.F.: The pathology of sudden deafness. Laryngoscope, St.Louis *72:* 1142–1147 (1962).

24 Schuknecht, H.F.; Kimura, R.S.; Naufal, P.M.: The pathology of sudden deafness. Acta oto-lar. *76:* 75–97 (1973).

25 Schuknecht, H.F.: Pathology of the ear (Harvard University Press, Cambridge 1974).

26 Veltri, R.W.; Wilson, W.R.; Sprinkle, P.M.; Rodman, S.M.; Kavesh, D.A.: The implication of viruses in idiopathic sudden hearing loss: primary infection or reactivation of latent viruses? Otolaryngol. Head Neck Surg. *89:* 137–141 (1981).

H.F. Schuknecht, MD, Department of Otolaryngology, Massachusetts Eye and Ear Infirmary, 243 Charles Street, Boston, MA 02114 (USA)

Adv. Oto-Rhino-Laryng., vol. 31, pp. 247–252 (Karger, Basel 1983)

Congenital Syphilitic Deafness –
A Long-Term Follow-Up

A. G. Kerr, D. A. Adams

Eye and Ear Clinic, Royal Victoria Hospital, Belfast, Northern Ireland

Introduction

Since 1967, 35 cases of congenital syphilitic labyrinthitis have been diagnosed in the Eye and Ear Clinic in Belfast. Some were bilaterally profoundly deaf on the first examination, some have been lost to follow-up and some have been diagnosed only in the past 6 years. This paper reviews those who had useful hearing in one or both ears at the time of diagnosis and who have been treated and followed up for at least 6 years.

It has proved impossible to find detailed statistics on the natural progression of the deafness of congenital syphilitic labyrinthitis. However, it is generally accepted that, untreated, there is fluctuation with gradual deterioration of the hearing towards the final state of bilateral profound deafness. Against this background, especially in an uncommon and slowly progressive condition, it has not been considered justifiable to do any controlled study. The goal of treatment has been the preservation of hearing; the avoidance of bilateral profound deafness has been looked upon as success.

Clinical Material

17 patients have been followed-up over a period of 6–14 ½ years, with a mean follow-up time of 10 ½ years. In earlier reports on this series, the patients have been divided into three groups – A, B and C.

Group A were those who were given ampicillin alone in a dose of 1.5 g four times daily for 4 weeks. There is now only 1 patient in this group (two ears). The other patients have now also been treated with prednisone or both prednisone and ACTH.

Table I. Hearing and speech discrimination scores (SD%) in patients in group B. 7 patients (14 ears)

At first attendance		At present	
	SD, %	SD, %	
4 dead ears	–	–	4 ears remain profoundly deaf
6 ears with fluctuating hearing	28	56 ⎫	2 ears, hearing now stable
	96	92 ⎭	
	52	64 ⎫	
	48	60 ⎪	4 ears, hearing continues to fluctuate
	56	60 ⎬	
	52	64 ⎭	
4 ears with stable hearing	92	88	now fluctuating
	36	16	hearing considered useless
	68	68 ⎫	2 remain stable
	92	92 ⎭	

Group B patients have been treated with ampicillin, 1.5 g 4 times daily for 4 weeks, and prednisone 10 mg 3 times daily for 10 days, gradually reducing over a subsequent 10 days. This group contains 7 patients (14 ears). Group C patients have had prolonged courses of ACTH in addition to treatment with ampicillin and prednisone. The most frequently used dose of ACTH has been 40 U intramuscularly each week but this has varied from 40 U per month to 120 per week. The dose has been adjusted in relation to the response of the patient. This group contains 6 patients (twelve ears).

A new group (D) has been added as 3 patients have died since the earlier classification. However, audiograms were available until shortly before their deaths and as they had been followed-up for nine, eleven and 13 ears it would have been unreasonable to have excluded them.

Throughout the follow-up all patients have had regular pure tone audiometry and assessment of speech discrimination. In considering the response to treatment, changes in speech discrimination have proved to be more meaningful than changes in pure tone responses.

Table II. Hearing and speech discrimination scores (SD%) in patients in group C. 6 patients (12 ears)

At first attendance		At present	
	SD, %	SD, %	
3 dead ears	–	–	3 ears remain profoundly deaf
8 ears with fluctuating hearing	84	88	1 ear, hearing now stable
	60	76	
	56	68	
	16	72	6 ears, hearing continues to fluctuate
	20	56	
	12	64	
	28	40	
	68	0	1 ear, now profoundly deaf
1 ear with stable hearing	72	84	hearing in this ear remains stable

Results

Group A. This patient had fluctuating hearing at the time of diagnosis and although his hearing has continued to fluctuate, his speech discrimination scores remain within normal limits and he is satisfied with his hearing.

Group B (table I). Of the 14 ears in this group, four were either profoundly deaf or unserviceable when the patient first attended. No response to treatment was expected or obtained in these ears. When the patients were first seen the hearing was fluctuating in 6 ears. This has persisted in four of these ears. The hearing in the other two ears is now stable. Useful hearing has been maintained in all six ears. There was no history of fluctuation in the hearing of the remaining four ears prior to treatment. In two of these ears the hearing has remained stable and speech discrimination scores are within normal limits. Fluctuation has now developed in the other two ears and in one the hearing has deteriorated to a level where it is considered unserviceable.

Group C (table II). Three of the twelve ears in this group were profoundly deaf when the patients first attended and, as expected, have remained so. In eight ears the hearing was fluctuating when the patients were first seen. In one

Table III. Hearing and speech discrimination scores (SD%) in patients in group D. 3 patients (6 ears)

At first attendance		At death	
	SD, %	SD, %	
3 dead ears	–	–	3 dead ears
3 ears with fluctuating hearing	48	56 ⎫	hearing continued to fluctuate
	36	32 ⎬	until death
	88	80 ⎭	

of these ears the hearing has become stable after treatment, with very good speech discrimination. In six ears the hearing still fluctuates but has been maintained at a useful level. In the remaining fluctuating ear the hearing was maintained, with gross fluctuation, for 11 years. However, during the past 3 years her hearing has gradually deteriorated despite high doses of steroids. The hearing in this ear is now essentially useless and ACTH is being withdrawn. Tragically, she had already lost all hearing in the other ear before treatment was started and she is now bilaterally profoundly deaf.

In the twelfth ear in this group the hearing was stable when the patient was first seen and has remained stable.

Group D (table III). 3 patients have died. All had been in group C and each had one profoundly deaf ear. In each the hearing of the remaining functioning ear was fluctuating prior to treatment and continued to fluctuate after treatment but useful hearing was maintained in each case until death.

Consequently, in this review of 14 living and 3 deceased patients, in whom there were 24 hearing ears, the hearing has been preserved at the pretreatment level or better in all but two ears (fig. 1).

Discussion

In the early cases in this series the patients were treated with ampicillin alone. If no response to treatment was observed, or if the improvement was not maintained, prednisone was also given. If, following this, deterioration occurred again, regular injections of ACTH were prescribed.

The relatively long period of follow-up of these patients allows an assessment of the efficacy of this treatment. There was optimism, initially, that mas-

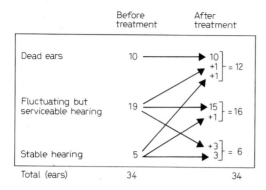

Fig. 1. Overall results.

sive doses of ampicillin would be adequate to control the disease. However, ampicillin alone does not control the progress of the disease in the majority of cases and there is now only 1 patient in group A. Although his hearing remains within normal limits it is still fluctuating and there must be some unease about his future.

In those ears which were profoundly deaf at the onset of treatment we have neither expected nor obtained any improvement in the hearing. However, of the ten serviceable ears in group B at the start of treatment, nine are still functioning satisfactorily and in only one is the hearing unserviceable. In group C there were nine serviceable ears at the start of treatment and there is serviceable hearing in all but one.

On the assumption that treatment did not shorten life, the 3 patients in group D can be considered successful. They were maintained with serviceable hearing until the times of their deaths.

In most of the publications on this subject, high doses of antibiotics have been recommended, often for long periods. *Kerr* et al. [1] have measured ampicillin levels in perilymph and found them satisfactory, but this was done only in non-syphilitic ears. It can be argued that because of the endarteritis, antibiotic levels in diseased ears could be lower than that required to kill the spirochaete. This may especially be true for the end stages of the disease where the blood supply to the cochlea is so poor that an antibiotic, no matter how given, may not reach bactericidal levels in the perilymph or the enchondral bone of the labyrinthine capsule.

Another potential drawback to the use of oral ampicillin is poor patient compliance. In this series, great care was taken to stress the importance of treatment to the patients and their family doctors. The patients were reviewed weekly during the treatment period and questioned closely about their medication. If any problems arose the patient was admitted to hospital for a period.

Although the action of the antibiotic is obvious, the action of the steroid is uncertain; this may well be the suppression of an immune response or simply an anti-inflammatory effect.

Vasodilators have been advocated by *Pilsbury and Shea* [3] to induce cochleo-vestibular blood flow, especially when the hearing is fluctuating. However, since syphilis causes an obliterative endarteritis, it seems unlikely that in advanced disease vasodilators would have much effect on the cochlear blood supply.

Conclusions

Ampicillin alone cannot be relied upon to control the progress of the disease in congenital syphilitic labyrinthitis. In combination with prednisone, massive doses of ampicillin have been effective in about half the cases. In the remainder, regular injections of ACTH have resulted in the maintenance of hearing in the majority of cases. Consequently, in 22 out of 24 hearing ears at the onset of treatment, useful hearing has been maintained over a period of 6–14½ years. Although detailed information is not available on the natural history of untreated cases the evidence suggests that, over this period of time, many more of these ears would have progressed to profound deafness. However, the fact that the hearing continues to fluctuate in many of these ears, still gives cause for concern.

References

1   Kerr, A.G; Smyth, G.D.L.; Landau, H.D.: Congenital syphilitic labyrinthitis. Archs Otolar. *91:* 474–478 (1970).
2   Kerr, A.G.; Smyth, G.D.L.; Cinnamond, M.J.: Congenital syphilitic deafness. J. Laryngol. Otol. *87:* 1–12 (1973).
3   Pilsbury, H.C.; Shea, J.J.: Luetic hydrops: diagnosis and therapy. Laryngoscope, St. Louis *89:* 1135–1144 (1979).
4   Smyth, G.D.L.; Kerr, A.G.; Cinnamond, M.J.: Deafness due to syphilis. J. Otolaryngol Soc. Aust. *4:* 36–40 (1976).

A.G. Kerr, MD, Eye and Ear Clinic, Royal Victoria Hospital,
Belfast (Northern Ireland)